SHEFFIELD HALLAM UNIVERSITY
LEARNING & IT SERVICES
ADSETTS CENTRE CITY CAMPUS
SHEFFIELD S1 1WB
WITHDRAWN FROM STOCK

SHEFFIELD HALLAM UNIVERSITY
LEARNING CENTRE
WITHDRAWN FROM STOCK

Kev

SHEFFIELD HALLAM UNIVERSITY

LEARNING CENTRE

WITHDRAWN FROM STOCK

SHEFFIELD HALLAM UNIVERSITY

LEARNING

WITHDR

Key Texts in Human Geography

Edited by
Phil Hubbard, Rob Kitchin and Gill Valentine

Los Angeles • London • New Delhi • Singapore

Editorial arrangement © Phil Hubbard, Rob Kitchin and Gill Valentine 2008

Chapter 1 © Bo Lenntrop
Chapter 2 © Michael F. Goodchild
Chapter 3 © Martin Charlton
Chapter 4 © Ron Johnston
Chapter 5 © Andy Wood
Chapter 6 © David Seamon and Jacob Sowers
Chapter 7 © Tim Cresswell
Chapter 8 © Noel Castree
Chapter 9 © Martin Phillips
Chapter 10 © Nick Phelps
Chapter 11 © Susan Hanson
Chapter 12 © David Gilbert
Chapter 13 © Satish Kumar
Chapter 14 © Jonathan Beaverstock

Chapter 15 © Keith Woodward and John Paul
  Jones III
Chapter 16 © Claudio Minca
Chapter 17 © Neil Coe
Chapter 18 © Nick Spedding
Chapter 19 © Robyn Longhurst
Chapter 20 © John Pickles
Chapter 21 © Phil Hubbard
Chapter 22 © Jo Sharp
Chapter 23 © Philip Kelly
Chapter 24 © Sarah Dyer
Chapter 25 © Alan Latham
Chapter 26 © Ben Anderson

First published 2008

Apart from any fair dealing for the purposes of research or private study, or criticism or review, as permitted under the Copyright, Designs and Patents Act, 1988, this publication may be reproduced, stored or transmitted in any form, or by any means, only with the prior permission in writing of the publishers, or in the case of reprographic reproduction, in accordance with the terms of licences issued by the Copyright Licensing Agency. Enquiries concerning reproduction outside those terms should be sent to the publishers.

SAGE Publications Ltd
1 Oliver's Yard
55 City Road
London EC1Y 1SP

SAGE Publications Inc.
2455 Teller Road
Thousand Oaks, California 91320

SAGE Publications India Pvt Ltd
B 1/I 1 Mohan Cooperative Industrial Area
Mathura Road
New Delhi 110 044

SAGE Publications Asia-Pacific Pte Ltd
33 Pekin Street #02-01
Far East Square
Singapore 048763

Library of Congress Control Number 2007940824

British Library Cataloguing in Publication data

A catalogue record for this book is available from the British Library

ISBN 978-1-4129-2260-9
ISBN 978-1-4129-2261-6 (pbk)

Typeset by C&M Digitals (P) Ltd., Chennai, India
Printed in Great Britain by TJ International Ltd, Padstow, Cornwall
Printed on paper from sustainable resources

SHEFFIELD HALLAM UNIVERSITY
WIL
304.2
KE
ADSETTS LEARNING CENTRE

Contents

Contributors

Ben Anderson is Lecturer in Human Geography, University of Durham, UK

Jonathan Beaverstock is Professor of Economic Geography, Nottingham University, UK

Noel Castree is Professor of Human Geography, University of Manchester, UK

Martin Charlton is Senior Research Fellow, National Centre for Geocomputation, National University of Ireland, Maynooth

Neil Coe is Senior Lecturer in Human Geography, University of Manchester, UK

Tim Cresswell is Professor of Human Geography, Royal Holloway, University of London, UK

Sarah Dyer is Lecturer in Human Geography, Oxford University, UK

David Gilbert is Professor of Urban and Historical Geography, Royal Holloway, University of London, UK

Michael F. Goodchild is Professor of Geography, University of California, Santa Barbara, US

Susan Hanson is Professor of Geography, Clark University, US

Phil Hubbard is Professor of Urban Social Geography, Loughborough University, UK

Ron Johnston is Professor of Human Geography, Bristol University, UK

John Paul Jones III is Professor of Geography, University of Arizona, US

Philip Kelly is Associate Professor of Geography, York University, Canada

Satish Kumar is Lecturer in Human Geography, Queen's University Belfast, UK

Alan Latham is Lecturer in Geography, University College London, UK

Bo Lenntrop is Emeritus Professor, Department of Geography, Stockholm University, Sweden

Robyn Longhurst is Professor of Geography, University of Waikato, New Zealand

Claudio Minca is Professor of Human Geography, Royal Holloway, University of London, UK

Nick Phelps is Reader, Bartlett School, University College London, UK

Martin Phillips is Reader in Social and Cultural Geography, University of Leicester, UK

John Pickles is Earl N. Phillips Distinguished Professor of International Studies, University of North Carolina, US

David Seamon is Professor of Architecture, Kansas State University, US

Jo Sharp is Senior Lecturer in Human Geography, University of Glasgow, UK

Jacob Sowers is a doctoral student in the Geography programme, Kansas State University, US

Nick Spedding is Lecturer of Geography, University of Aberdeen, UK

Andy Wood is Associate Professor of Human Geography, University of Kentucky, US

Keith Woodward is Lecturer in Human Geography, University of Exeter, UK

Acknowledgements

The editors want to express their gratitude to all the contributors for responding so positively to their invitation to contribute to this volume, and for working to tight deadlines. We also wish to thank Robert Rojek for his patience and encouragement whilst we bought this project to completion.

Among our contributors, Susan Hanson wishes to acknowledge discussions with Sophie Bowlby and Megan Cope on Geography and Gender. Ron Johnston wishes to thank Les Hepple, Tony Hoare, Kelvyn Jones, Charles Pattie and Eric Pawson for valuable discussions of his essay and comments on draft versions. Nick Phelps would like to thank Doreen Massey, Philip Cooke and Mick Dunford for providing further background on the radical economic geographical scene of the late 1970s and early 1980s. Satish Kumar acknowledges assistance from Niall Majury, Nuala Johnson, Diarmid Finnegan and Harjit Singh in rehearsing the account of capitalism, neo-liberalism, and dialectics that underpins his chapter, and also thanks David Zou for assistance in bibliographic compilation. Phil Hubbard also wishes to thank all members of the Loughborough University Department of Geography reading group for sharing their thoughts on David Sibley's *Geographies of Exclusion*.

© 26.1 'Ceci n'est pas l'espace' Massey, Doreen. *For Space*. SAGE, London 2005.

List of Figures and Tables

Figures

Tables

Editors' Introduction

Phil Hubbard, Rob Kitchin and Gill Valentine

Why key texts?

Geography, like all academic disciplines, is never static, with geographers always seeking to either extend and consolidate particular ways of thinking and doing or to develop new understandings of the unfolding relationship between people, place and environment. Far from being a discipline preoccupied with the mere accumulation of facts about the world, geography is a discipline where our understanding of the world is constantly being evaluated in the light of new ideas and thinking, with empirical projects always informed by notions that some forms of knowledge and ways of knowing may be more productive or valid than others. Empirical studies of what appears to be happening in particular contexts thus build up into wider theoretical accounts that, in turn, drive new explorations of how people, place and environment are entwined in complex and relational geographies. Without this diversity of thought and sense of progression – i.e. the idea that we are moving towards a more productive understanding of the way the world works – geography would long ago have become an intellectual backwater, rather than the vibrant, vital and varied discipline that many currently believe it to be.

This book is based on the premise that texts play a crucial role in this story of disciplinary development. More specifically, it works with the assumption that particular texts can be read and interpreted as symptomatic (and perhaps totemic) of key transitions in the ways that we think, practise and write geography. The widest possible definition of a geography text might include conference papers, journal articles, book chapters, literature reviews, working papers, online articles, monographs, student textbooks, dictionaries, encyclopedias, readers, gazeteers, maps, and atlases. All of these may, in different ways, presage important shifts in the way geography is conducted. Yet in this volume we want to focus on one particular kind of text: the book. More specifically we want to hone in one type of book – the authored monograph.

Despite some concern in the discipline that publishers are less and less willing to publish monographs, preferring instead to publish student-oriented texts like this one, we focus here on authored monographs for two principal reasons. First, while most monographs are empirical – in the sense they seek to describe or map a particular aspect of the world – many also make important theoretical statements about the way that geographical knowledges should be constructed and disseminated. For many, this is the acid test of a geographic research monograph: for a book to make a contribution that furthers the discipline, it needs to spell out possible routes towards a more relevant, ethical or viable geography by advocating a particular approach to its subject matter. As such, monographs seek to transform geographical thinking and praxis through a sustained engagement with, and exploration of, a set of theoretical ideas as well as the detailing of particular empirical 'facts'. They are often books with Big Ideas and Big Ambitions, and if their thesis gains currency they become key reference works which are mined, re-worked and critiqued by subsequent generations. Authored texts thus become foils that stimulate new ideas and thinking. This is not to say that research articles, chapters, edited books, etc. do not make similarly important contributions to debates concerning disciplinary progress. Indeed, papers in journals may often stimulate important changes in the disciplinary landscape by providing rapid dissemination of research findings. Yet we would argue that authored books often become *the* key milestones in disciplinary histories in ways that articles rarely do because they allow authors to connect disparate empirical and theoretical elements to develop a wider, more systematic and rigorous argument about the way the world works.

Second, we focus on authored monographs because students are often referred to these 'key texts', encouraged to engage with them in order to understand particular modes of thought and the history of the discipline and to reflect on the ideas contained within them with respect to shaping their own geographical thinking and praxis. Many courses on the histories and philosophies of geography are in fact stories in which key authored texts are given due prominence, with these key works deemed to have punctuated geography's histories. From the perspective of the present, a retrospective reading of these works is often encouraged as a way of understanding how we got to where we are today. Making oneself familiar with key texts is part of any geographical education – for many educators, 'thinking geographically' is something that only emerges from critical reading and re-reading of geography's 'ur-texts'. Contemporary libraries, we note, are often all-too-ready to dispose of older books to make way for new tomes, but most retain those volumes which educators suggest are 'classics' which students will always need to return to.

This brief discussion indicates that this volume is necessarily restrictive in its definition of what a 'key text' is. Not only do we ignore papers, chapters, edited collections, readers and conference presentations, we also disregard a number of important student-oriented textbooks. Ron Johnston (2006) argues that textbooks are particularly important in institutionalizing particular approaches to geography, given that they often proclaim to be 'objective' or authoritative introductions to the discipline. Although we would not necessarily disagree, given that such volumes are written to be accessible to as wide an audience as possible, we feel there is little need to provide a guide to such texts. Nor do we consider some of the important histories of the discipline (Johnston's, 2005, *Geography and Geographers* – now in its fifth edition – being a prime example, alongside Peet's, 1998, *Modern Geographical Thought,* or Cloke, Philo and Sadler's, 1991, *Approaching Human Geography*) for the same reason. Other books that have gone through multiple iterations, and have hence been integral in policing the boundaries of the discipline, are also precluded from consideration here (e.g. the *Dictionary of Human Geography*, now in its fifth edition, the *Companion Encyclopedia of Human Geography*, now in its second, the *International Encyclopedia of Human Geography*, and so on).

By focusing on specific authored texts we are accordingly not trying to suggest other types of text are insignificant in shaping geographical thought. Yet we feel that, by their very nature, the type of books we focus on here were often written for an academic peer audience rather than a student one. For the uninitiated, many appear remarkably dense, use difficult language and work through complex theory and unfamiliar examples, and are not accessible in the same way that a textbook might be. Many talk to debates and social-economic contexts that have long since disappeared or are only just coming into being, or refer to events or processes that would only be known to those who have grown up in particular contexts. In a sense, this is precisely why lecturers are often keen to refer their students to these texts – they want to challenge them, encourage them to develop their skills of critical reading and appreciate the impact of particular thinkers and ways of thinking on the practices of geography.

This given, *Key Texts* provides an introduction to 26 books that we argue have made a significant impact on the theoretical underpinnings and praxis of human geography in the last 50 or so years. The book's ambition is two-fold. First, it aims to serve as a primer for students, introducing them to specific monographs, exploring the nuance of the authors' arguments and explaining why they should take the time and trouble to engage with the text itself rather than summaries provided in textbooks. To that end, each of the entries in this volume is an interpretive essay that highlights: the positionality and biography of the author(s); the significance of the text in relation to the geographical debates and issues current at the time of writing; the book's main arguments and sources of evidence; its initial impacts and reception; how the book was subsequently critiqued, evaluated and incorporated into the geographical imagination; and how the book changed – and continues to influence – the practices of geography.

Secondly, the book seeks to contribute to ongoing debates over the production of geographic knowledge by posing some important questions of what constitutes a 'key text'. It is of course crucial to ask why some books become privileged, and to consider how disciplinary histories become written around key texts as well as key thinkers (see Hubbard *et al.*, 2004). In recent years scholars interested in the history of geographical knowledge production have come to argue that geographical endeavour occurs in a highly diversified landscape, shaped by issues such as educational training,

personality and location, friendships and collegiality, disciplinary gatekeeping and access to disciplinary networks, prevalent trends and vested interests, and wider debates on the relevance and value of the academy and the funding of higher education. In other words, it has become recognized that geographic scholarship is shaped by multiple factors, some personal, cultural and social, and some that are more political and economic in nature (Barnes, 2002). While the academy is a place of collegiality and collaboration, it can also be a competitive environment with most academics working both for themselves and their institutions as they seek to acquire kudos, funding and intellectual respect. In the UK, for example, departments are in competition with one another under the influence of a Research Assessment Exercise which is focused on research outputs. The influence of RAE culture on the shape and form of institutional geography is still to become clear, yet the potential to be identified by one's institution as a 'research inactive' academic creates immense pressures to work in particular ways, and to work to identified assessment criteria.

These diverse factors shape what kinds of ideas and praxis become mainstream, and, in turn, influence who become recognized as the key thinkers in a discipline (though, as the exchange in *Environment and Planning A* 37: 161–187, illustrates, there are certain dangers in trying to name those who are most influential in the discipline). However, as debates about English language and Anglo-American dominance in the production of geographic knowledge have highlighted, knowledge production has both a history and a geography, with some scholars located in key centres, others on the periphery (see Berg and Kearns, 1998; Garcia-Ramon, 2003; Kitchin, 2005; Paasi, 2005). As such, it is important to acknowledge that the production of geoographical knowledge is messy, contingent, relational and political, meaning that any history of the discipline needs to be written in ways that are cognizant of such politics. *Key Texts* is no exception.

The authors of the entries in this volume were not asked to explicitly address such issues, but often raise questions of authority, privilege and hierarchy in their contributions. As many acknowledge, it is seldom the case that a book becomes significant because of sheer good fortune. Rather, a book catches and helps further promote the zeitgeist of a particular moment, crystallizing the authors' thoughts at the time, and often many other peoples', working within and outside the discipline. Indeed, it is likely that many conference papers and journal articles preceded the publication of a 'key text', especially given how long books take to write. And it is often the case that other related or similar books appeared at roughly the same time. What distinguishes a 'key text' – a book that was taken up or 'enrolled' within dominant disciplinary networks – is that it said something significant that had widespread appeal and which challenged its readers to think differently about the world. It simply did not repeat arguments emerging within the journal literature, it extended these, amplified them and illustrated them with rich empirical material. Of course, other books might have been saying similar things, but were perhaps saying them less well, with less conviction or were promoting slightly different viewpoints. And while the book's content is the crucial factor determining its reception and uptake, there is no denying that issues of authorship and authority are also significant. In short, it matters who wrote the book. Some books are highly anticipated given their author's existing reputation; others emanate from unheralded sources but become best-selling works. Most, however, are published to indifference, and never achieve anything more than modest sales: the ability of an author to promote their books through their other activities and networking can be vital in ensuring a book has a shelf-life.

What is clear is that some books emerge to become 'classics' within the discipline. The authors of such books may (reluctantly or otherwise) become gatekeepers within the discipline – recognized as 'key thinkers' – in the sense both they and their books are held up as promoting a particular way of doing geography. As such, many of the entries in this volume highlight interesting debates about the politics of geographical knowledge production. So too does our choice of key texts. In making the difficult choices we have made about which books are worthy of consideration, we realize that we are not simply reflecting established knowledges, we are actively perpetuating particular value claims about whose views matter, and which books should be read. That given, we are certain that the books we exclude will be as significant for some readers as the books we include – and we hope that these exclusions are interrogated as meaningfully and productively as was the case for the companion work for this text – *Key Thinkers on Space and Place* (see especially the review forum in *Environment and Planning A* 37: 161–187). Given the controversy that our selection will undoubtedly excite in some quarters, it is hence necessary to spend some little time outlining the criteria for selection that we have employed in this volume.

Which key texts?

When drawing up a list of some of the most important texts in human geography, we are forced to make some difficult decisions as to what we understand the boundaries of human geography to be. Indeed, even if we are happy to exclude key texts in physical geography (the subject of a volume yet to come?), there are certainly books on the relation of the physical and human world that have been significant in changing the purview of geographers and their understanding of what the subject matter of the discipline might be. There is also the vexed question of what distinguishes a geographical text from other kinds of text, given many key interventions in debates over space and place have been made by those who do not identify as geographers or claim to be writing for a geographical audience. The boundaries between human geography and planning, urban studies, history, anthropology, sociology, philosophy and area studies have often been highly porous and geographical thinking has certainly borrowed and benefited significantly from texts written by those located in other disciplines.

Given these fluid and indistinct boundaries, our first criterion for selection was to consider only books written by people who self-identified as geographers, and were writing, first and foremost, for a geographical audience. This is to take a narrow view of the discipline, perhaps, but is in keeping with one of the most widely accepted (if hoary) definitions of geography: that is, geography is what geographers do. Secondly, we limited our choices to books published in English in the last 50 years. While this is a somewhat arbitrary cut-off date, there are good reasons for supposing that students will be most frequently steered towards these texts: many recent, student-friendly histories of the discipline tend to start with the post-war shift from regional description to a theoretically inclined spatial science tradition (i.e. what is commonly referred to as the 'quantitative revolution'); students are most likely to be steered to books whose ideas still have currency in contemporary debates (and these tend to be the most recently published); many university libraries simply do not have an extensive catalogue of books dating back to the pre-World War Two years, and students undertaking courses in Anglophone countries

often lack an advanced proficiency in languages other than English that would prevent them from engaging with non-English texts.

Within these broad parameters, we still faced difficult decisions about what constituted a 'key text' and which 'key texts' to include. One strategy to aid our selection might have been to consult the citation rates for different books (i.e. the number of times a book has been referred to in other books and articles). There is some tradition of using citation analysis to identify the 'weavers and makers' of human geography (e.g. Bodman, 1991), and online databases and search engines (Google Scholar™ or the ISI Indexes) certainly facilitate such analyses. Yet not all such analyses are robust and reliable, and we should remain mindful that not all citations are favourable. Equally, some books appear to be more cited than read (a charge made in at least one of the chapters in this volume), and self-citation can often inflate the apparent importance of a given text. Suffice to say, most of the books in this volume are well-cited (some more so than others), but not all of the most cited books in geography are included here.

Another way of honing in on the key texts within the discipline might have been to select the best-selling texts. However, for a variety of reasons, sales might not be a good indicator of significance. As we noted above, textbooks, especially those designed for 'introduction to human geography' courses, tend to have significantly more sales than research monographs. This is because monographs are generally targeted at the author's peer group rather than student masses, and so might have limited sales. They might, however, have hundreds more citations than a textbook which sold in far higher numbers (suggesting that they sway more influence). Further, geography books seldom (if ever) break into the best-seller charts, with few geographers ever having adopted the position of a genuine 'public intellectual' (see Ward, 2006; Castree, 2006 on public geographies). Some texts may well achieve sales beyond geography, reaching student and academic audiences in cognate disciplines, but very few break through to 'discerning' public audiences in the same way that, for example, historical or archaeological books currently do in the UK. This does not mean that the key texts we profile have not sold well, with some featured here having gone to multiple editions and repeat print runs. Irrespective of this, we would claim that the acid test of a key text is not its popularity or citationality, but its longevity; that is to say that their impact is best measured in terms of their influence on subsequent texts.

While mindful of citations, sales, and longevity, we chose to narrow our selection further by consulting with colleagues from across human geography as to what books they felt merited inclusion based on their experience as researchers and teachers. From that extended list we whittled the books down to the publisher's limit of approximately 25 (given page length constraints). Here, we tried to provide a regular temporal spacing of books, including texts published within each decade; to include texts that were important within specific sub-disciplines as well as human geography as a whole; and to include texts that engaged with and promoted the many '-ologies' and '-isms' that have permeated recent geographical thought. We also took the decision to try to include some quite recent books that we feel have the potential to become 'key texts' given their initial reception and how quickly their ideas have permeated the discipline.

This then is not a random selection of books. It is a set of books that we believe are worth reading, either individually or collectively. Each book has made an important intervention not just within a given sub-discipline (e.g. urban, rural, social, economic, political, historical or cultural geography) but shaped the wider practices and imaginations of human geography. Indeed, if one were to critically read all the books included in this volume, one would have a very good grasp of geographical theory and practice over the past 50 years. It is nonetheless a subjectively derived list and we would in no way claim it is *the* list of the most influential books in human geography. Other geographers would have of course drawn up their own lists and may be distraught to find some of their favourite texts excluded here (possibly including their own books!) They will no doubt suggest that our list bears the imprint of our own particular exposure to geography through Anglo-American traditions, our own research interests and expertise, our own peer networks – all of which might have impinged on our judgement as to which books have exercised most influence on geographical thought. This is unavoidable, not least because any book published prior to 1990 pre-dates our professional experience in the discipline (indeed the earliest books were written before all three of us were born!). But, for all its flaws, we feel that the 26 entries in this book do some kind of justice to the diversity of human geography practised in the last 50 years, with each having 'pushed the envelope' intellectually, methodologically and philosophically, shaping the landscape of human geography as we see it today.

How to use this book

The most important thing to stress about this book is that it is *not* intended to substitute for an engagement with the text itself. While each entry provides a synopsis of the book in hand, it is necessarily brief, and often glosses over the nuance of the book's argument in the interests of identifying its essential arguments. The book is designed to be a primer, to be read *alongside* the text, and seeks to encourage a critically informed engagement and exploration of the intricacies of each book. Each of our entries thus provides useful background context that might help the reader understand the situation within which a particular text was written (i.e. the wider social and political conditions that prevailed at the time, as well as the disciplinary preoccupations which prompted the authorship of that volume). It also considers how the book was received by the wider academic community, noting the way that the book was reviewed and how people reacted to the ideas being forwarded. In many cases, the entry also documents how people engaged with the ideas and took them forward in different ways. Accordingly, each entry considers the way in which the book affected and shaped the geographies that succeeded it, through a critical appraisal of the book's key thematic concerns, its particular approach and espousal of specific philosophies of geographic knowledge production.

Those asked to write chapters for this volume were asked to do so because we felt they would be able to offer a critical, reflective and balanced assessment of the text in question. Inevitably, many of our authors have written about a book that has profoundly shaped their own life as a geographer, perhaps influencing their own approach to the discipline. Some are explicit about this, and provide a highly personalized account of how the book influenced them; others are less forthcoming, and instead try to produce an account in which their own opinions are harder to discern. But in either case it is highly unlikely that the authors have produced an unbiased interpretation, with most likely to be predisposed towards (or, occasionally, against) the book they are considering. This is unavoidable: as we have stressed, there is no such thing as an objective assessment, and there is no one who is in a position to ultimately determine the value of a text. What therefore needs to be remembered is that each of our entries comes with a point of view that you,

and your tutors or colleagues, might not share.

Our hope is thus that this book will prove a useful companion for students seeking to engage with geography's rich histories (and not a crib that will dispense with the need to actually read each book in question) and might provide a useful template for how students might critically engage with texts in general (indeed a useful exercise is for students to undertake critical reviews of texts not included in this volume). The bibliography that closes each entry consequently offers numerous departure points for further explorations of the text and its author, and throughout we include frequent cross-references to other entries in this volume (as well as the companion text, *Key Thinkers on Space and Place*). As we suggested above, when we commissioned people to write the entries in this book, we chose people who we felt might have a close knowledge or perhaps even affinity for the book in question. Each is also knowledgable about the place of the book in the wider disciplinary landscape and in the 'geographical tradition'. Yet we implore students not to take their views for granted, as their summation is not necessarily the one that other geographers might make. Perhaps their reading contradicts your own, or comes to different conclusions. Any one of the books featured here is open to multiple readings, and sometimes even readings that the author never intended. Such is the polysemous nature of text. In the final analysis, we hope that this book provokes people to read and re-read these texts, subject them to discussion and interrogation, and form their own situated interpretations. Perhaps, in time, these engagements might even stimulate the production of new key texts! Whatever, we hope *Key Texts* is a useful and stimulating text that is much, much more than a lesson in geographical navel-gazing and nostalgia.

Secondary sources and references

Barnes, T. (2002) 'Performing economic geography: two men, two books and a cast of thousands', *Environment and Planning A* 34: 487–512.

Berg, L.D. and Kearns, R.A. (1998) 'America unlimited', *Environment and Planning D: Society and Space* 16: 128–132.

Bodman, A. (1991) 'Weavers of influence: the structure of contemporary geographic research', *Transactions, Institute of British Geographers* 16: 21–37.

Castree, N. (2006) 'Geography's new public intellectuals', *Antipode* 38: 396–412.

Garcia-Ramon, M.D. (2003) 'Globalization and international geography: the questions of languages and scholarly traditions', *Progress in Human Geography* 27: 1–5.

Hubbard, P., Kitchin, R.M. and Valentine, G. (2004) *Key Thinkers on Space and Place*. London: Sage.

Johnston, R.J. (2006) 'The politics of changing human geography's agenda: textbooks and the representation of increasing diversity', *Transactions, Institute of British Geographers* 31: 286–303.

Kitchin, R. (2005) 'Disrupting and destabilizing Anglo-American English-language hegemony in geography', *Social & Cultural Geography* 6: 1–15.

Paasi, A. (2005) 'Globalization, academic capitalism and the uneven geographies of international journal publishing spaces', *Environment and Planning A* 37: 769–789.

Review Forum: 'Key Thinkers on Space and Place', *Environment and Planning A* 37: 161–187.

Ward, K. (2006) 'Geography and public policy: towards public geographies', *Progress in Human Geography* 30: 495–503.

1 INNOVATION DIFFUSION AS SPATIAL PROCESS (1953): TÖRSTEN HÄGERSTRAND

Bo Lenntrop

The diffusion of innovations – the origin and dissemination of cultural novelties – is an area of study which concerns all sciences dealing with human activity, including, not least of all, cultural and economic geography. (Hägerstrand, 1953: 1)

Introduction

It is difficult to grasp the importance of Törsten Hägerstrand's key work on innovation diffusion – his doctoral thesis from 1953 – without an appreciation of the historical context in which the work was conceived and prepared and the fact that it was first translated into English by Allan Pred some 14 years after submission. Most notable for setting out theories of spatial diffusion and adoption, Hägerstrand's early research also contains the key to the later development of his ideas. One example is *time geography*, which became formalized in the 1960s, but whose conceptual roots were to be found in Hägerstrand's writing in the 1940s and 1950s.

Hägerstrand arrived at Lund University in the late 1930s. His interest was directed, more or less by chance, toward migration and he started to work on a project intended to chart the entire demographic development of a considerable geographical area of Sweden from 1840 to 1940. Building an impressive collection of data, this enormous undertaking left Hägerstrand with a profound empirical understanding of demographic development.

At the same time he also developed a deeper theoretical proficiency, setting in motion his particular geographical worldview. More particularly, the systematic collection and analysis of data on the life courses of a population over a century contributed to the germination and growth of a foundational idea; namely the importance of analysing spatial processes.

Over time Hägerstrand absorbed important theoretic ideas and trends from beyond Swedish geography. These did not emanate from the regional perspective then dominant in university teaching, with Hägerstrand stating 'lectures in regional geography were abominably boring … Geography appeared not as a realm of ideas or a perspective on the world but as an endless array of encyclopaedic data' (Hägerstrand, 1983: 244). Rather they came to his attention through a chance acquaintance. His future wife, Britt, was then working for the ethnologist Sigrid Svensson who was conducting research and publishing books on diffusion processes, and one of his school colleagues had a burning interest in numerical analyses and in developing computers and was an early visitor to the US.

Having gained a sound knowledge of demography and ideas on diffusion courses and simulation models, the foundations of his doctoral thesis had fallen into place. In many respects this work was to mark a decisive break with the then dominant tradition of regional studies. Hägerstrand's principal aim in his doctoral thesis was not to present a broad regional description of an area, but instead to investigate and illuminate a problem. That the material concerned a specific area, argued Hägerstrand, was a regrettable necessity and not a methodological finesse which in itself marked a stand against regional geography.

Innovation diffusion: the Swedish version

For obvious reasons a doctoral thesis written in Swedish does not reach a large international readership. Nevertheless, it is informative to comment on its immediate reception, 14 years before it was first published in English. The first academic review of Hägerstrand's doctoral thesis was published in 1953 – in the same year he was awarded his doctorate. The reviewer was Edgar Kant, previously professor in Tartu in Estonia, but who had by then spent many years in Lund at the Department of Geography. Kant was widely read in international scholarly literature and he was also the first opponent at Hägerstrand's disputation.

Kant opens his review with a discussion of the respective pros and cons of research specialization. Kant was of the opinion that the disadvantages associated with specialization 'begin to become apparent when large lacunae arise leaving poorly-lit areas in border zones, as monadnocks of the total ignorance' (Kant, 1953: 221). It is useful to recite Kant's concluding comments in order to relate exactly why he saw Hägerstrand's thesis as pioneering.

The author has, to a noteworthy extent, utilized new methods and established new links to neighbouring disciplines. This must present itself as innovative to those who perceive geography as indissolubly bound to traditional methods of investigation and research subjects, such as landscape analysis, which have as their only or primary task ascertaining interactions between man and nature. ... It may transpire that the author's longest expeditions into the unknown border- and twilight-zone have been but excursions leaving many areas as yet largely unexplored. Those who follow in his path can, however, draw benefit from his pioneering work and fashion new riches. (Kant, 1953: 225)

In his doctoral thesis Hägerstrand investigated the changing extent of propagation of cultural artefacts. He did this by selecting six specific indicators; three for agriculture (state subsidies for improving pastures, control of bovine tuberculosis, soil mapping) and three more general indicators (postal money transfer, automobiles, telephones). The choice of indicators was guided by the need that these should be localizable to coordinates and that their development over time could be followed with a very high degree of precision. Furthermore, it was necessary that the indicators had been adopted by a sizeable proportion of the population.

The next stage in the investigation was to establish reduction bases for the indicators. This was necessary because it would have been meaningless to work with absolute numbers of acceptors. Hägerstrand carefully analysed the population (or demographic) development in the region and the location and size of each of the respective homesteads and residential apartments in order to construct the 'reduction bases' against which the number of acceptors should be matched.

An important section of the thesis, encompassing nearly 100 pages, then deals with the actual diffusion process. Hägerstrand's primary objective here is not to identify specific details in this diffusion process, but rather to generalize about the characteristics that could serve as the basis for subsequent operationalization in models. However, it was important to observe certain characteristics as these generated variations in the diffusion process. A significant degree of the discrepancy between the indicators is based on the level of state intervention, for example in promoting the introduction of controls on bovine tuberculosis. Car purchase was principally a matter of private decision making, even if state legislation played a certain role. The introduction and diffusion of the telephone in the region of investigation was affected by the manner and the speed with which the electricity network was developed. Consequently, the diffusion of the six indicators reveals different courses and show how they are influenced to lesser or greater extents by planning and policy at national and regional levels. The larger part of this discussion is of more general interest, and is not confined to its historical or regional context.

With the help of this detailed empirical knowledge, Hägerstrand then formulated a series of experimental, stochastic models to show how innovations spread within a population. The first model was very simple and is steered entirely by chance and provides a picture that most closely approximates the manner in which a rumour spreads through a population. Hägerstrand therefore focuses on how differences in acceptance and how unevenly spread information could be modelled. Concerning the indicators for agriculture, he examines how, for example, farm size can influence the propensity to accept an innovation. In his studies, Hägerstrand identifies the importance of the proximity and thus comes to deal with private information diffusion and the question of how this should be modelled. He examines the chorological characteristics of information and how migration and telephone data can be utilized to describe the extent of the range of private information. In this particular case study he finds that migration data provides the best approximation as the telephone network was incomplete at the time of the study.

Furthermore, there were zonal boundaries that acted to deform the contact field.

The resulting models operated with a real, coordinate based population. The diffusion of an innovation in a population is determined by constructing a so-called mean information field (a concept still outlined in many standard textbooks on human geography), which illustrates how the probability of making contact with another individual decreases with increased distance. The empirical basis for the MIFs comprising 5×5 cells (each cell being 1 square kilometre) is grounded on migration data. The matrix illustrates the probability of a contact being made from the central cell to one or more of the surrounding cells. The probabilities are cumulated (from 1 to 10,000) and each cell is attributed an interval proportional to the probability. The matrix is centred on an individual having knowledge of the innovation in question. By drawing a random number one decides which cell (interval) is met. The matrix is used in the manner of a floating grid – i.e. it moves over the fixed population and is centred over those individuals who have knowledge of the innovation in each generation and who are prepared to spread this knowledge. In this manner the innovation is continually diffused over time to new generations and gives rise to spatial patterns of acceptors, which are randomly determined but always within the given probability intervals. Even when the rules of the game and the probabilities remain unchanged, the results of different simulations are often very divergent as a result of the stochastic factor.

In these models Hägerstrand experimented with different forms of physical and social barriers. The propensity to accept an innovation was modelled (for example, a person must be 'hit' two or three times before accepting) and the degree of correspondence with the empirically ascertained patterns became increasingly close. As a result of the degree of correspondence between the actual and the modelled courses, Hägerstrand was able to conclude that the key elements deciding the course of innovation diffusion had been captured. It is worth noting that all of these comprehensive modelling experiments were carried out by hand as computers had not as yet been developed. Each of the thousand upon thousand of random numbers was collated from tables in a strict order.

The first international commentary on Hägerstrand's doctoral thesis was written by John Leighly and published in the *Geographical Review* in 1954. Leighly, then Professor at Berkeley, was well acquainted with Swedish geography, which he had considered for several years (Leighly, 1952). Leighly emphasized his admiration of the scope and precision of the empirical work, not least concerning the map representation: 'His "relative" mapping uses refinements (isarithms at numerical intervals given by geometric progressions, interpolation of isarithms by logarithmic intervals) that make it exemplary' (Leighly, 1954: 440). However, and in a similar manner to many other commentators, Leighly viewed Hägerstrand's interpretation of innovation diffusion and the following operational modelling as the 'culmination' of his work. Leighly (1954: 441) concludes by commenting that anyone doing research in this area 'cannot afford to ignore Hägerstrand's methods and conclusions'.

While the attention surrounding Hägerstrand's doctoral thesis waned in the period following this appraisal, the same cannot be said regarding interest in his theoretical ideas and methodological approach. Aspects of his principal research were gradually disseminated via lectures, conferences and minor publications. Shortly prior to the completion of his doctoral thesis, Hägerstrand laid out his research field in an article entitled 'The propagation of innovation waves' (Hägerstrand, 1952). By 1967 he had published three articles on migration, two on diffusion, and one on simulation. A very widely read and often cited article is 'A Monte Carlo approach to diffusion' (1965), which

directly deals with models and is published in four journals and has even been translated into Japanese.

Innovation diffusion: the English version

Although Hägerstrand's research had become relatively well known on the international scene, the broader understanding of his research remained fragmentary. Accordingly, both Gilbert White in Chicago and Allan Pred at Berkeley argued that his doctoral thesis should be translated into English. Pred took on this task as he was both fluent in Swedish and well versed in the specific area of research; two attributes which Hägerstrand argued were prerequisites for the successful translation of his work. Pred also wrote a Postscript in which he introduced Torsten Hägerstrand providing a detailed background to the research, and how the field of research had developed.

The book was well received. However, the fact that 14 years had passed since the publication of the original Swedish edition is at least partly evidenced by the content of a number of more critical commentaries. Quantitative geography had further developed and was quickly establishing an important set of new methodologies and theories (see Chapters 2, 3 and 4). One particular line of development concerned the statistical and objective comparison of patterns in time and space, which made it possible to elucidate more precisely differences and similarities in these patterns. In his review, Gunnar Olsson pointed out that 'the evaluation of the model was based on more intuition and visual inspection than objective statistical testing' (Olsson, 1969: 310). None of the reviews published during the 1950s had raised this issue.

Simulation had become a popular *modus operandi* within the discipline, and against this background Olsson had somewhat critically stated: 'In conclusion, most of those who now experiment with Monte Carlo models seem to have missed the point. Thus it is rare that the relation between theory and model is as explicit as in Hägerstrand's own work' (Olsson, 1969: 311). This observation further supports the proficiency with which Hägerstrand had succeeded in fusing thorough empirical material with a well-developed command of theory. This is what gives his work its credibility and strength.

Richard L. Morrill, himself a former visiting fellow at Lund who also had experience of simulations, argued that although Hägerstrand's earlier work had been disseminated relatively widely, the opportunity now existed for a wider audience to discover that this was by no means a curious approach, but instead constituted a theoretical framework founded on an unusually rigorous empirical grounding. Morrill (1969) went to particular lengths to emphasize the linkages between spatial patterns and individual behaviour, and argued this represented a revolutionizing insight and a powerful break with tradition within the discipline. In a more anecdotal vein, we might also note that he concluded his review in *Economic Geography* by protesting against the 'outrageous price' of the book (then $16).

L.J. Evenden (1969) forwarded a comprehensive and positive review in the journal *Social Forces*. He argued that: '… Hägerstrand allows a glimpse of a fertile imagination and, simultaneously, teaches the useful lessons that patience is a research virtue, that theory is fundamentally about *something*, and that reliable theory in social science stays close to this *something*' (Evenden, 1969: 9). Evenden further argued that evidence that Hägerstrand's research had left a lasting legacy was to be found in the wide range of fields in which Monte Carlo simulation had subsequently been applied. Simulation was well known and widely used at the time of the

publication of the English edition, so there was no question of its subject matter being received as breaking news. Evenden emphasized that the work represented a combination of a careful factual analysis and the best of traditional and contemporary geographical scholarship, and contended that Hägerstrand's doctoral thesis would long be considered a classic work in the development of geographical theory.

The economist Harvey Leibenstein, then based at Harvard University, reviewed the book in the *Journal of Economic Literature*. He maintained that it was of interest for economists who were interested in diffusion processes and location theory, as well as modulated processing of innovation diffusion. Leibenstein further contended that it was difficult to assess the value of the models' predictions, in spite of all the empirical information. He also found the approach altogether mechanical – a critique which one can at least partly agree upon – if one did not take into consideration the fact that the translated thesis was 14 years old. A further point made by Leibenstein was that it had been a mistake to translate the thesis in its entirety. The meticulous account of empirical material was, he deemed, utterly boring and would probably act to '… repel most readers' (Leibenstein, 1969: 11). It is as well to add that no such critique has been forthcoming from any of the geographers who have reviewed the book; instead they have viewed this detail as an extra bonus – stimulating for many, previously unknown for most readers.

As time goes by

While Hägerstrand's research on diffusion was largely well known prior to the publication of the English version, it did open the possibility for a deeper understanding of his research. At first it was widely read and acted to inspire many researchers, but one gains the impression that it fairly quickly became a work that was deferentially cited for its voluminous length and that the main elements of his research were largely disseminated through his articles. Another contributing reason was that Hägerstrand did not further develop his diffusion research. He abandoned research on diffusion courses at a relatively early stage in his career. However, two-thirds of the references to his work in academic journals concern this field of research. The English version of his doctoral thesis, with almost 400 citations, is by far the most cited.

There is still much to be gained from reading Hägerstrand's doctoral thesis, particularly the manner in which he constructed his models. It was his models and the use of Monte Carlo-simulation that had the greatest impact and these resulted in citations and references and acted as the models for many subsequent variants. In a number of articles one can find abridged discussions concerning fields of private information and the influence of proximity in the diffusion of ideas. A reading of Hägerstrand's doctoral thesis does, however, provide much more – his reasoning on propagation courses, on planning and politics relating to diffusion processes, etc. The reader will also perceive the consequential interplay between the empirical and the theoretical.

Without an awareness of the doctoral thesis, it would be easy to disregard the importance of empirical data for advanced idea construction and pioneering research. The importance of compiling a base of knowledge from fundamental empirical work, with methodological and theoretical thinking being intimately connected, is perhaps the most important lesson to be learned from Hägerstrand's research on diffusion. Assuredly, for the contemporary readers, familiar with shorter journal articles and student-friendly

texts, Hägerstrand's doctoral thesis is dauntingly comprehensive. However, it is well worth reading as a classical scientific work and one which is also a *tour de force* in terms of map and diagram presentation.

In his doctoral thesis Hägerstrand worked with models that were operationalized by hand, but all along these were envisaged with future computer technology in mind. He was cognisant of the important role computer technology would play, not least for geographers, and in an article (in Swedish) published in the mid-1950s, he pointed out areas of application for coordinate based information and how this could be analysed. This later inspired Hägerstrand's early involvement with areas of application such as coordinate based real estate registers and the development of computer mapping; a forerunner to GIS.

Viewed after a great number of years, Hägerstrand's writing from the 1940s and 50s retains its vitality and relevance. His book was presented in *Progress in Human Geography* as one of the discipline's classic works (1992), with comments by Andrew Cliff and Allan Pred and an authorial response by Hägerstrand himself. His work in the field of diffusion research has also been cited in *The Dictionary of Human Geography* where it is referred to at length.

In his portrait of Hägerstrand, Flowerdew (2004) sketches a broader picture of his research that does not confine itself to diffusion. However, he names this area of research as one of Hägerstrand's more important contributions, arguing that it included much in the way of new thinking and that it greatly influenced the wider research community. The same spirit, but expressed in other shades of meaning, is conveyed in the image forwarded by many researchers of Hägerstrand as one of the founders of modern human geography (*Progress in Human Geography*, 2005). For interested readers a newly published bibliography is available (Lenntrop, 2004).

Conclusion

At an early stage in his research Hägerstrand emphasized proximity – in both geographical and social networks – as a crucial factor in accessing and spreading ideas (not least emphasized in time-geography). Through his contacts, which also partly arose by chance – he gained, as we have seen examples of above, new building blocks for his research. Absorbing new impulses and ideas from outside is, however, insufficient without being able to choose and then fuse together new and older ideas into new combinations and constructs. Hägerstrand had both the ability and a world view with exactly these qualities. It was by no means a statistical understanding but a dynamic one, which after years of development also had roots reaching long back in time.

A further trademark of Hägerstrand's research was a focus on processes. In his doctoral thesis he rejected the enterprise of comparing conditions at different moments in time and instead endeavoured to directly elucidate the processes that formed different spatial patterns. He grasped reality as interacting processes in which events and actions of one moment conditioned events and actions of the next and not as variables related to each other (Hägerstrand, 1992). This reasoning came to influence the entire construction of time-geography and its associated conceptual apparatus.

Despite the copious amounts of time and energy that Hägerstrand invested in collecting and interpreting empirical material, he always retained the notion of generalization and the aim of bringing forth the underlying factors rather than losing one's way in the surface details. This is a trademark for his entire body of research and can easily be traced in his work. With the aid of a few concrete examples he was able to show fundamental relations or paths of development.

Secondary sources and references

Evenden, L.J. (1969) 'Review of Innovation Diffusion as a Spatial Process by Törsten Hägerstrand', *Social Forces* 47 (3): 356–357.

Flowerdew, R. (2004) 'Törsten Hägerstrand', in P. Hubbard, R. Kitchin and G. Valentine (eds) *Key Thinkers on Space and Place*. London: Sage, pp. 149–154.

Hägerstrand, T. (1952) 'The propagation of innovation waves', *Lund Studies in Geography*, Ser B. Human Geography No. 4.

Hägerstrand, T. (1953) Innovationsförloppet ut korologisk synpunkt. Akad.avh. *Meddelande från Lunds Universitets Geografiska Institution. Avhandlingar* XXV. Lund 1953.

Hägerstrand, T. (1967) *Innovation Diffusion as a Spatial Process*. Translation and Postscript by Allan Pred. Chicago and London: University of Chicago Press.

Hägerstrand, T. (1983) 'In search for the sources of concepts', in A. Buttimer (ed.) *The Practice of Geography*. Harlow: Longman, pp. 238–256.

Hägerstrand, T. (1992) 'Author's response. In Classics in Human Geography revisited', *Progress in Human Geography* 16 (4): 543–544.

Kant, E. (1953) Recension av Törsten Hägerstrand Innovationsförloppet ur korologisk synpunkt. *Svensk Geografisk Årsbok*, Årg. 29: 222–225.

Leibenstein, H. (1969) 'Review of Innovation Diffusion as a Spatial Process by Törsten Hägerstrand; Allan Pred', *Journal of Economic Literature* 7 (4): 1213–1214.

Leighly, J. (1952) 'Population and settlement: Some recent Swedish studies', *Geographical Review* 42 (1): 134–137.

Leighly, J. (1954) 'Innovation and area under geographical record', *Geographical Review* 44 (3): 439–441.

Lenntrop, B. (2004) 'Publications by Törsten Hägerstrand 1938–2004', *Geografiska Annaler, Serie B, Human Geography*, 86 B (4). Updated version on www.keg.lu.se/ Publikationer/Törsten Hägerstrand.

Morrill, R.L. (1969) 'Review of Innovation Diffusion as a Spatial Process by Törsten Hägerstrand', *Economic Geography* 45 (3): 356–357.

Olsson, G. (1969) 'Review of Innovation Diffusion as a Spatial Process by Törsten Hägerstrand; Allan Pred', *Geographical Review* 59 (2): 309–311.

Progress in Human Geography (2005) 'Makers of modern human geography – Törsten Hägerstrand (1916–2004)', *Progress in Human Geography* 29 (3): 327–349.

2 THEORETICAL GEOGRAPHY (1962): WILLIAM BUNGE

Michael F. Goodchild

All geographers have essentially the same problem, to make sense out of the globe. (Bunge, 1966: xvii)

Introduction

Theoretical Geography first appeared in 1962, and in an expanded second edition in 1966 (which is the volume cited here). To Cox (2001: 71) it is 'perhaps the seminal text of the spatial-quantitative revolution. Certainly in terms of laying out the philosophical pre-suppositions of that movement it had no peer.' But Cox goes further:

> It was also the spatial-quantitative revolution that gave impetus to conceptual precision in the field ... it was the prospect of measurement, of operationalization in some piece of empirical research, that helped us discover the value of a careful specification of our concepts and an examination of their consistency ... So, if we want to see where we have come from, what our intellectual debts are, there are few better places to start than *Theoretical Geography*. (Cox, 2001: 71)

The rationale for the book, laid out in the Introduction, is that geography is a science; that every science is defined by its domain of knowledge, which for geography is the Earth as the home of humanity; that every science has both a factual or empirical side and a theoretical side; and that 'there are many books

on geographic facts and none on theory' (Bunge, 1966: x). Moreover, given 'the author is a theoretical geographer' (Bunge, 1966: x) – theory is what he does. Theory 'must meet certain standards including clarity, simplicity, generality, and accuracy' (Bunge, 1966: 2), and only through theory can we discover 'these patterns, these morphological laws ... so that (our) planet, Earth, fills (our) consciousness with its symmetry and ordered beauty' (Bunge, 1966: xvi).

The book and its contents

The medieval Christian idea of a geometrically perfect world centered on Jerusalem (Harley and Woodward, 1987) had been abandoned in the Renaissance, and replaced with a view that the Earth's surface was infinitely complex. So, although science had made great strides in explaining other aspects of the cosmos, at the same time, by apparently rejecting religious teaching, it had led to a confusing view of human behavior, and of our immediate surroundings. In Bunge's view, it was only through geographic theory that we could return our concepts of the world to an emotionally satisfying and reassuring simplicity.

For Bunge, there were simple laws to be discovered about the Earth's surface, and particularly about the patterns of phenomena found there. They would be found for both human and physical phenomena, and because the study of one could inform the other it made sense for departments of geography to house both physical geographers and human geographers. The key to understanding patterns was geometry, which allowed the precise description of pattern and, through its theorems, allowed the researcher to reason about pattern and thence to theory.

The book opens with a chapter on the methodology of geography which makes the author's position clear. Two examples are used to focus attention on what will be the recurring geometric theme of the book. Regional geography, long dominated by descriptive work that emphasized the unique properties of places, could be put firmly within a scientific paradigm if one focused on devising objective and replicable methods for defining regions, and concentrated on statistical evidence instead of subjective impression. As a second example, the shifting river courses of the lower Mississippi bore a striking geometric resemblance to the historic development of north-south highways in the Seattle area, and similar explanations could be found if the geomorphological process of building natural levees were compared with the social process of commercial development along arterial roads.

The second chapter, titled 'Metacartography,' takes another icon of traditional geography, the map, and goes to great lengths to argue its essential mathematical nature, and the degree to which such core cartographic concepts as projection, overlay, and generalization can be expressed in the mathematical frameworks of geometry and set theory. A chapter on shape discusses potential measures of this somewhat elusive property, while subsequent chapters cover sampling, topology, and geodesics. A full chapter on central place theory reflects the early 1960's focus on this as a crucial element of geography's 'theoretical turn' (see Kelly, Chapter 23, this volume). The book hints at many of the major developments in quantitative geography and geographic information

science over the coming decades, though of course it is impossible to find anticipation of today's emphasis on computation and computer networks.

But there is far more to the book than is suggested by this quick summary, as one might infer from the commentary from Cox quoted above. The book needs to be understood within the context of debates that raged in geography at that time, and in the context of the much wider intellectual movement of which it is in many ways the opening statement. It also needs to be understood within the ageless desire of geographers to rank among the respected disciplines of the academy.

Debates within geography

To Bunge, the most important intellectual debate in geography in the late 1950s and early 1960s was that between the *nomothetic* and *idiographic* perspectives on science. Although their actual positions were much more nuanced, geographers often identify the perspectives with the names of Schaefer (1953) and Hartshorne (1959) respectively. Bunge had been strongly influenced by Schaefer, having been drawn to his work by William Garrison while a student at the University of Washington, and it is no accident that the first citation in the book is to his paper.

The nomothetic position holds that results in science are of value only when they are general, applying equally at all locations in space and time. For example, Mendeleev's Periodic Table would be far less significant if the elements behaved differently during leap years, and Newton's Laws of Motion would be far less significant if they applied only in the State of Minnesota. In other words, empirical science proceeds by abstracting and generalizing away from the space-time context in which all observations are made; and scientific results are applied by re-inserting them into context.

For Bunge science was inherently nomothetic, and geography was a science. Its success would be measured strictly by the number of general principles that it discovered, within its defined domain of knowledge. These would be principles governing the patterns that emerge on the Earth's surface as a result of social and physical processes, ranging from the dendritic patterns of stream channels to the regular spacing of settlements. The purpose of geography was to discover these principles, and the chapters of his book laid the foundation, by outlining a methodology and providing examples.

Geographers are clearly going to find this nomothetic position difficult, since the description of variation over the surface of the planet is so much part of the intellectual tradition of the discipline, and much would be lost if description were justified only if it led to general principles. So while physicists might marginalize *mere* description, geographers must necessarily treat it with more respect. The idiographic position holds that description has inherent value, particularly if it is conducted according to scientific principles – for example, using generally agreed terms, and methods that are described in sufficient detail to be replicable by others. But Bunge and others rejected this position, arguing that an idiographic geography would always be a marginal science.

To Bunge, then, it was important that geography establish its credentials as a science. The first edition appeared in 1962, at the very outset of what later became known as the *quantitative revolution* in geography (Berry, 1993). Bunge was one of a group of graduate students who had been inspired at the University of Washington in the late 1950s by William Garrison, Edward Ullman, and Donald Hudson. The group included Duane Marble, Brian Berry, Michael Dacey, Richard Morrill, John Nystuen, Arthur Getis, and Waldo Tobler. As these fresh PhDs gained positions in influential universities in the US, their approach rapidly caught the imagination of others,

including Leslie King, Leslie Curry, Maurice Yeates, Peter Haggett and Richard Chorley in the UK, among others. Further key texts appeared, including Haggett's *Locational Analysis in Human Geography* (1966) (see Charlton, Chapter 3, this volume), Chorley's and Haggett's *Models in Geography* (1967), and King's *Statistical Analysis in Geography* (1969). New journals were established, positions created in almost all departments, and by the end of the decade it could truly be said that geography had been transformed.

But it would be oversimplifying to suggest that this transformation merely altered the balance between nomothetic and idiographic visions of science. As Cox argued in the passage quoted earlier, Bunge also contributed to a transformation of geographic practice that included all aspects of the scientific enterprise. It came at a critical point, when computers were becoming available as engines of analysis and modelling. His book makes a passionate case for mathematics – the *lingua franca* of science – with its well-defined and widely accepted terms and its formal reasoning, and for particular applications of geometry to the analysis of geographic pattern. It argues the merits of quantification, and of the formal reasoning of statistical inference; moreover, as we have seen, it is at pains to elaborate the relationship between maps, the traditional research tool of the geographer, and mathematics, the pervasive tool of science.

Reading the book again after 40 years, one is struck with the sense of excitement of the early quantifiers, and by a rhetorical style that clearly emulates the sciences and is rarely encountered in geography today. In the introduction to the second edition, published in 1966, Bunge writes:

the discovery of a great number of patterns of location during the fall of 1962 was helpful. We now know why political units have their overall shape, why rivers are dendritic, and why alluvial fans and Burgess's rings are such close spatial cousins. Also, the problem of the construction of the isometric surface is much better understood. (Bunge, 1962: xiv)

Nevertheless the author cannot resist continuing:

To see region construction, one of the last preserves of the non or anti-mathematical geographers, crumble away before the ever growing appetite of the computing machines is a little unnerving even for a hard case quantifier. (Bunge, 1962: xiv)

The revolution was well on the way to victory.

Counter-revolution

Where as *Theoretical Geography* may indeed have been the 'shot heard round the world,' and while it provided a general outline of how to make the pitchforks, guns, and guillotines that would be needed before the revolutionaries could prevail, much of its content does not stand up well to detailed analysis. While it argues strongly for mathematics, there is precious little actual mathematics in the book, and it fell to others such as Alan Wilson (1970) to provide the rigorous theorizing. It is easy to see parallels between the behavior of floating magnets in a dish of water and the behavior of vendors positioning themselves on a featureless plain (Bunge, 1962: 283). But on more careful analysis it is obvious that the laws of magnetic force do not have precise analogs in the laws of entrepreneurial behavior. Bunge was (and still is) the passionate visionary, but it took others with a deeper knowledge of mathematics and a patient commitment to rigor to begin to build the peer-reviewed literature of quantitative geography. Moreover, it was not long before the counter-revolution began.

One of the first salvos was fired by Sack (1972) (see also Sack, 1973; Bunge, 1979). Certainly one could find consistent patterns across the Earth's surface – the geometric form of the river meander, Horton's laws of stream number (Horton, 1945), the regular spacing of settlements in Southern Germany – but geometric pattern in and of itself could not be regarded as *explanation*, and could not therefore provide the sense of satisfaction that comes from knowing how a pattern is caused. Spatial properties such as latitude, distance, or direction could never be said to truly explain anything. Sack saw Bunge as the primary exponent of a confusion between geometric pattern and explanation, and attacked his notion that it was an interest in space and spatial properties that defined the discipline.

This issue remains an important one today. The spatial tradition is still a key part of the discipline, though only one of many in a pluralistic field (Pattison, 1964). At the same time, few would claim today, as Bunge seems to have done in 1962, that geography somehow has or should have a monopoly on all things spatial. Sack points out that it is not distance itself that explains human behavior, but the costs or other impediments that result from spatial separation, such as transport cost or the time taken to travel. But if these are approximated by mathematical functions of distance this argument seems to be hairsplitting – to all intents and purposes it is distance that explains why people have more friends nearby than far away, even in the Internet age, and why fresh milk is more likely to come from nearby cows.

Moreover the floating magnet example points to another fundamental problem, the fact that many different processes can lead to similar geometric forms. Thus it can be difficult to deduce general statements about process from the patterns found on the Earth's surface, particularly if one cannot observe how those patterns change through time. The spatial clustering of cases of a disease provides a compelling example. In the traditional re-telling of the story, the map made by Dr John Snow of the cholera outbreak in London in 1854 showed a clear clustering around the pump that he had suspected of providing cholera-carrying water, and thus of causing the outbreak (see, for example, Longley *et al.*, 2005). But a cluster in space can always arise through two very different processes. In a first-order process the likelihood of an event is a function of location, and events are more likely in some locations than others, in this case near the pump. In a second-order process events affect each other, and the presence of one event makes others more likely in the immediate vicinity; an example in epidemiology would be contagion, in which a disease is passed directly from one individual to another. Under the hypothesis of a contagious process, which at the time was how cholera was generally believed to have been transmitted, an initial carrier of the disease who happened to live near the pump could have caused the same spatial pattern. Only by observing the pattern's development through time could one acquire evidence to resolve between the two hypotheses.

The book's lasting impact

It was problems and issues such as these that led geographers in the 1970s to question the value of Bunge's heavy reliance on the spatial approach, and to look elsewhere for more productive sources of explanation. Many others remained firmly committed, however, and continued to refine the methods and models that Bunge had pioneered. Computing provided an additional impetus, as it became possible to bring much more powerful tools to bear on geographic data. By the 1980s geographic information systems (GIS) had become a focus for much of this work, and many of Bunge's ideas had been implemented and successfully applied. The rather simplistic ideas of central place theory had evolved into location-allocation, his discussion of optimal routing and geodesics had blossomed into corridor location, and his primitive ideas of spatial similarity had spawned metrics of spatial autocorrelation and the fields of spatial statistics and geostatistics.

Today, Bunge's ideas of a geography grounded in geometry are alive and well, but not for quite the reasons he suggested. As many have argued (see, for example, Laudan, 1996), science is not only about the rather emotional topic of explanation, but also about prediction, design, and other useful human activities. Goodchild and Janelle (2004) list six arguments for a spatial approach in the social sciences (and note that all six apply to both space and time):

- **Integration**. A spatial approach allows information to be placed in context, and allows information from different sources to be correlated. These are often presented as the most compelling functions of GIS.
- **Cross-sectional analysis**. As Bunge argued, spatial patterns can often provide insight into process, particularly if patterns are observed through time.
- **Spatial theory**. A spatial theory can be defined as one that includes such spatial properties as location, distance, or direction in its structure. Central place theory, theories of spatial diffusion, and theories of spatial ecology all have this property.
- **Place-based analysis**. These are recently developed methods of analysis that recognize the fundamental variability of the Earth's surface, and produce maps of results rather than single summary statistics. The geographically weighted regression devised by Fotheringham *et al.* (2004) is a good example.
- **Prediction, design, and policy formulation**. As noted earlier, the application of

general results requires that they be placed back into a space-time context. To predict the effects of a change in oil prices on the economy of California, for example, it is necessary to take general results on how economies respond to oil prices and to apply them in the specific space-time context of California.

- **Information storage and retrieval**. Space and time provide a powerful way of organizing knowledge, and of retrieving information from today's vast stores of digital data (see, for example, the Alexandria Digital Library, www.alexandria.ucsb.edu, and the overview of *geoportals* by Maguire and Longley, 2005).

And what of the author himself? William Bunge is a very imposing person, two meters in height and powerfully built, and with enormous energy. Like many others, he was strongly engaged by the political turmoil of the 1960s, and frustrated by the slow, reflective pace of academic life and what he perceived as its lack of direct influence and action. He felt a strong urge to demonstrate the practical importance of the ideas he espoused in *Theoretical Geography*. As a faculty member at Wayne State University, he spent several years implementing his ideas in the ghettos of Detroit (Bunge, 1971), while continuing to participate in the meetings of the Michigan Inter-University Community of Mathematical Geographers (MICMOG), which were held close to the point of minimum aggregate travel of the University of Michigan, Michigan State University, and Wayne State University (in a high school in Brighton, MI). The Detroit Expedition became a seminal exercise in working with disadvantaged communities, applying many of the ideas of the early quantifiers towards the improvement of the human condition. Detailed maps of human deprivation were created based on observation, and the group worked aggressively towards the adoption of better school districting plans that

implemented many of the ideas of location theory.

Bunge took a strong stance against the Vietnam War, and became increasingly disillusioned with both US society and US academe. His positions led to increasing friction both on and off campus, and in the early 1970s he left the US for Canada, where he took short-term positions at the University of Western Ontario and York University. The graduate seminar he gave at the University of Western Ontario was well received by many of the students, but his openly expressed disgust with the political positions of some of his colleagues made it impossible to renew his contract. He drove taxis for a time in Toronto, and eventually settled in small-town Quebec. In 1988 Bunge's *Nuclear War Atlas* appeared (Bunge, 1988), a powerful application of spatial techniques to draw attention to the impossible consequences of the use of thermonuclear devices. The evolution and politicization of his thinking is clearly evident in his 1979 retrospective on *Theoretical Geography* that was published in the *Annals of the Association of American Geographers* to mark the AAG's 75th Anniversary (Bunge, 1979). The passion is there, more strongly than ever, as is his commitment to scientific method, empirical observation, and sound theory, but it is directed at different targets.

Conclusion

As Macmillan argued in another commentary (Macmillan, 2001), despite all the critiques of quantification and positivism that have appeared in the past three decades, Bunge's *Theoretical Geography* remains as 'a major landmark in the history of geographic thought' (1966: 74). It appeared 'on the cusp between the old world and the new' (ibid.: 74), between the old analog world of crude, imprecise tools and the modern world of abundant data and powerful techniques of analysis, visualization, and simulation. Reading the book one gets a sense of what an effort it must have been to prepare the maps, tables, and summary statistics using the pens, slide-rules, hand calculators, and log tables of 1962, and how much more one could have achieved today. Yet virtually all of the major areas of today's spatial perspective are there, in a book largely written while Bunge was a graduate student. There can be few periods in the history of any discipline when a group of people created quite the intellectual ferment that must have existed in geography at the University of Washington in the late 1950s and early 1960s. *Theoretical Geography* captured that ferment in a way that is still meaningful today.

Secondary sources and references

Berry, B.J.L. (1993) 'Geography's quantitative revolution: initial conditions, 1954–1960. A personal memoir', *Urban Geography* 14 (5): 434–441.

Bunge, W. (1962) *Theoretical Geography* (1st edn). Lund Studies in Geography Series C: General and Mathematical Geography. Lund, Sweden: Gleerup.

Bunge, W. (1966) *Theoretical Geography* (1st edn). Lund Studies in Geography Series C: General and Mathematical Geography. Lund, Sweden: Gleerup.

Bunge, W. (1966) *Theoretical Geography* (second edition) Lund Studies in Geography Series C: General and Mathematical Geography, No. 1. Lund, Sweden: Gleerup.

Bunge, W. (1971) *Fitzgerald; Geography of a Revolution*. Cambridge, MA: Schenkman.

Bunge, W. (1973) 'Commentary: spatial prediction', *Annals of the Association of American Geographers* 63 (4): 566–568.

Bunge, W. (1979) 'Perspective on *Theoretical Geography*', *Annals of the Association of American Geographers* 69: 169–174.

Bunge, W. (1988) *The Nuclear War Atlas*. New York: Blackwell.

Chorly, R.J. and Haggen, P. (eds). (1967) *Models in Geography*. London: Methuen.

Cox, K.R. (2001) 'Classics in human geography revisited: Bunge, W., Theoretical Geography. Commentary 1', *Progress in Human Geography* 25 (1): 71–73.

Fotheringham, A.S., Brunsdon, C. and Charlton, M. (2004) *Geographically Weighted Regression*. Hoboken, NJ: Wiley.

Goodchild, M.F. and Janelle, D.G. (2004) 'Thinking spatially in the social sciences', In M.F. Goodchild and D.G. Janelle, editors, *Spatially Integrated Social Science*. New York: Oxford University Press, pp. 3–22.

Haggett, P. (1966) *Locational Analysis in Human Geography*. London: St Martin's Press.

Harley, J.B. and Woodward, D. (1987) *The History of Cartography. Volume 1: Cartography in Prehistoric, Ancient, and Medieval Europe and the Mediterranean*. Chicago: University of Chicago Press.

Hartshorne, R. (1959) *Perspective on the Nature of Geography*. Chicago: Rand McNally.

Horton, R.E. (1945) 'Erosional development of streams and their drainage basins: hydrophysical approach to quantitative morphology', *Geological Society of America Bulletin* 56: 275–370.

King, L.J. (1969) *Statistical Analysis in Geography*. Englewood Cliffs, NJ: Prentice-Hall.

Laudan, L. (1996) *Beyond Positivism and Relativism: Theory, Method, and Evidence*. Boulder, CO: Westview Press.

Longley, P.A., Goodchild, M.F., Maguire, D.J. and Rhind, D.W. (2005) *Geographic Information Systems and Science*. New York: Wiley.

Macmillan, W. (2001) 'Commentary 2: geography as geometry', *Progress in Human Geography* 25 (1): 73–75.

Maguire, D.J. and Longley, P.A. (2005) 'The emergence of geoportals and their role in spatial data infrastructures', *Computers, Environment and Urban Systems* 29: 3–14.

Pattison, W.D. (1964) 'The four traditions of geography', *Journal of Geography* 63 (5): 211–216.

Sack, R.D. (1972) 'Geography, geometry and prediction', *Annals of the Association of American Geographers* 62: 61–78.

Sack, R.D. (1973) 'Comment in reply', *Annals of the Association of American Geographers* 63 (4): 568–569.

Schaefer, F. (1953) 'Exceptionalism in geography: a methodological examination', *Annals of the Association of American Geographers* 43: 226–249.

Wilson, A.G. (1970) *Entropy in Urban and Regional Modelling*. London: Pion.

3 LOCATIONAL ANALYSIS IN HUMAN GEOGRAPHY (1965): PETER HAGGETT

Martin Charlton

Locational analysis is concerned with the need to look for pattern and order in geography ... with the locational systems we study and the models we create to describe them, and with the types of explanation we use in making sense of our findings. (Haggett, 1965: 1–2)

Introduction

By 1945 most universities in Britain possessed a geography department. However, whilst many disciplines taught at university had a strong and rigorous intellectual coherence, particularly the sciences, geography was not one of them. Sydney Wooldridge and Gordon East, Professors of Geography at King's College and Birkbeck College London, had attempted to suggest some new directions for the discipline in *The Spirit and Purpose of Geography* (Wooldridge and East, 1951). Much research, however, was regional in nature, and human geography concerned itself with the identification of regions, their classification and description. Papers with titles such as *Horticultural Developments in East Yorkshire, Urban Regions of St Albans,* or *Egypt's Population Problem* were the rule rather than the exception. It has been suggested the Second World War was won with Science; that said, the technological leaps which had been made between 1939 and 1946 were not mirrored by equivalent progress in geography. By comparison with mathematics or operations research,

human geography's largely descriptive orientation was hardly scientific in its approach to understanding problems, and in particular, the problems of a world attempting to regain some semblance of normality after a conflict which had been global in nature.

There was, however, a scientific interest in geographical problems outside the discipline – as early as 1934 a small group of sociologists met in the Adelphi Hotel, Philadelphia, to discuss the statistical problems of using data for census tracts. Gehlke and Biehl (1934) and Neprash (1934) were, in this sense, way ahead of geographers. It would be another 30 years before geographers started to attack such issues effectively. Indeed, Trevor Barnes (2003) suggests that 1955 marks the date when one could point to the emergence of what has become known as spatial science in the US with activity at the Universities of Washington and Iowa, and subsequently, Chicago, Northwestern, Michigan, and Ohio. Barnes (2003) also suggests that the roots of this activity go back to earlier traditions, notably the work of von Thünen, Alfred Weber, and August Lösch. All three were searching for rationalism. Lösch's text, *The Economics of Location*, appeared in 1954. Barnes (2003) comments that Lösch's influence was strongest on those viewing geography as nomothetic (law-seeking) locational analysis rather than an idiographic (descriptive) view of regional economies. By 1960 a formidable range of graduates had emerged in US universities, including Brian

Berry, William Bunge, Michael Dacey, Arthur Getis, Duane Marble, Richard Morrill, John Nystuen and Waldo Tobler. Armed with an array of techniques and the computer, they were ready to change human geography. They had Science on their side.

On the other side of the Atlantic was Törsten Hägerstrand, Professor of Geography at the University of Lund. He too was interested in using the power of mathematics to model innovation diffusion. *Innovation Diffusion as Spatial Process* appeared in 1953, but was not published in English until 1967. The tool that Hägerstrand is credited with introducing to the social scientists is that of Monte Carlo simulation (Evenden, 1969). By contrast, little was taking place in Britain, but things were about to change. In 1957 Peter Haggett was appointed University Demonstrator in Geography at the University of Cambridge. Aware of the developments taking place in both the USA and Sweden he began to put together some lecture notes for undergraduates taking geography at Cambridge. As often happens with such collections a suggestion was made that the material might usefully be turned into a book. Haggett credits his colleague and co-author on many books and papers, Richard Chorley, with the idea. Haggett notes ruefully in his preface 'the manuscript was written painfully slowly'. Elsewhere there was little obvious move towards a theoretical geography: the contents of the *Transactions of the Institute of British Geographers* (IBG) changed little in the decade after the Swedish publication of Hägerstrand's volume.

In 1965, Haggett's *Locational Analysis in Human Geography* was published. An intellectual *tour-de-force*, it was unlike any other British text on geography. Later that year many IBG members must have been astounded on opening the December issue of *Transactions of the Institute of British Geographers* to find Stan Gregory's report on a symposium held at Bristol University on regression techniques in geography (Gregory, 1965), with a further

20 pages from Chorley and Haggett on trend surface mapping in geographical research (Chorley and Haggett, 1965). If those across the Atlantic needed evidence that human geography in the ivory towers of British academe was at last shifting out of the doldrums, this was it.

The book and its arguments

The published text stands at just over 300 pages, divided into two main sections preceded by a chapter entitled *Assumptions*. Part One, *Models of Locational Structure*, includes five chapters each with a single-word title: Movement, Networks, Nodes, Hierarchies, and Surfaces. Part Two, *Methods in Locational Analysis*, has four chapters, Collecting, Description, Region-Building, and Testing. Without the context set by the First Part, the Second Part would lack the theoretical structure which holds it all together.

In Part One, Haggett brings together much of the theoretical work in geography which had appeared in the literature from the middle of the nineteenth century onwards. He examines briefly geographers' search for order and their attempts to make sense of the world. He considers the nature of geography 'a thorn in the side of school and university administrators' (1965: 9). The emergence of regional science created a linkage between economics and geography which was the catalyst for an interdisciplinary approach to locational problems which was 'immensely productive'. One of the activities which emerges is model-building – 'to codify what has gone before and excite fresh enquiry'. The models which Haggett examines in Part One are for the most part normative – that is they describe some ideal state of spatial organization, although they are generally based on observations of human spatial behaviour. There are four principal theorists from whose work he draws: Johann von Thünen, Walter Christaller, Alfred Weber, and August

Lösch – all representing German geographical traditions. He later draws on more recent work from the United States: Michael Dacey, William Garrison, William Bunge, Walter Isard and Brian Berry are all heavily cited. There are apparently fewer figureheads from Europe and elsewhere, although Törsten Hägerstrand's work is dealt with in some detail.

Christaller's Central Place model appears in Movement (Chapter 2), Nodes (4) and Hierarchies (5). Weber's work is also cited almost exclusively in Hierarchies (5). Losch's contributions appear in all chapters in Part One. Among the 20th century commentators, Movements (2) has additional material from Hägerstrand, Isard and Bunge, Networks (3) cites Garrison and Bunge. Nodes (4) features work by Berry and Hierarchies (5) draws heavily on Isard's work. Von Thünen's work is covered extensively in Surfaces (6) with additional material from Sauer, Berry and Bunge. What Haggett achieves in Part One is to demonstrate the relationships between the individual components of locational analysis.

In Movement (2), we shift from movement along paths, and the associated idea of a least-cost path, through the modelling of interaction processes with gravity models, continuous movement across a field (which looks forward to Tomlin's map algebra), movement in zones, and finally to the temporal aspect of movement in which Hägerstrand's diffusion models are explained. Chapter 3, Networks, considers routes and routing. Some of the concepts – the least-cost path between two or more settlements – has diffused well outside geography. There are references to graph theory in mathematics but the connection with Dijkstra's shortest-path algorithm is tantalizingly absent, and discussion of concepts such as route density and network change.

Chapters 4 and 5, Nodes and Hierarchies, deal with settlements and their interrelationships. We start with models of settlement pattern, examing the deviations from regularity due to agglomeration, variations in resource location, and temporal change. Haggett examines the relationship between the size of a town and its rank in a list of towns in a region – the rank/size relationship and its various models. This is expanded into a discussion of the relationship between the size and spacing of settlements and what the appropriate models might be, and with respect to hierachies, their function and potentional distortions from models due to agglomeration or resource localization.

The final chapter considers Surfaces. Models may be of continuous processes, such as population density, or discrete processes, which are illustrated by considering land use zones. Thünian models of minimum-movement are given an extensive treatment – Haggett also links Weberian models for locating points in space with von Thünen's problem of locating areas in space. The models may be subject to distortion – the discussion leads us through Burgess' and Hoyt's models of city structure to a consideration of what happens when people do not behave as the models might suggest: the 'spectre of sub-optimal behaviour'. The chapter and the first part of the book end with the suggestion that models are not always appropriate, Haggett arguing that the model building should follow satisficer rather than optimizer principles. As Haggett notes ruefully, 'Most of our existing models have been with us far too long' (1965: 182).

In contrast, Part Two is informed more exclusively by geographical research of the time and seeks to place some logic on geographical methods – 'the ways in which geographical information can be gathered, measured, classified, and described … in order that our existing concepts may be critically examined' (1965: 185). Haggett starts with the amassing of evidence and proceeds towards the testing of hypotheses in these four chapters.

These chapters are not a cookbook of techniques borrowed from other disciplines, but the reader is assumed to have some basic statistical knowledge. What becomes clear on reading these chapters are the strides made since the early 1960s in the science of handling geographical data. In 1965 techniques and methods in quantitative geography were in their infancy or had yet to be developed, and the computing context, which we take for granted, was absent. Haggett hence starts the section with a discussion of issues of data collection. Many of the foci are still pertinent today – where do we find data? what sort of data might we find? and what might its characteristics be? Haggett moves gently from the abstract (what is a population?) towards the operational (how do we recognize our sample and the areas to which it relates?) He considers various approaches to sampling, illustrating the discussion with various different sample designs. He also considers the geographical element – noting that the data we collect are often for irregular or inconvenient or incompatiable spatial units – and provides examples of what would now refer to as areal aggregation or interpolation.

Sound advice given in many introductory quantitative analysis courses is that an analyst should explore data long before any modelling is carried out. Haggett's journey continues with Description. His approach is that the task is to 'describe briefly and accurately locational patterns revealed by data collection' (1965: 211). The basic tools are cartographic and statistical. Mapping data is the starting point describing 'absolute location'. In many ways the discussion seems somewhat basic but Haggett was writing before the development of interpolation techniques or geostatistics and representing several variables was as difficult then as it is now. Haggett considers some basic techniques before examining principal components analysis, albeit rather briefly. Again, we must remember that carrying out such an analysis was a major computational task. The geometer returns as Haggett then describes statistical indices for shapes, points, and networks.

In spite of the dominance of the region in geographical writing, much of Chapter 9 on Region-building will be unfamiliar to today's students. The underlying questions are those of creating regions grouping them. For example, what sort of characteristics do regions have? Consideration of this question leads into set theory, boundary superimposition and a collection of quantitative methods based on geometry. An important problem in areal geography is the modifiable areal unit problem which occurs when data are aggregated at different spatial scales, producing very different distributions, depending on how the areas are defined and mapped. Haggett calls this the scale problem and considers it at length, noting that geographers had not made much contribution to its analysis. He suggests methods to tackle it based on variance decomposition, filter mapping, and trend surface modelling.

Today's student would recognize much of the final chapter – Testing. In Haggett's view testing 'provides an appropriate check on the theoretical excesses of Part One' and presents an 'enragingly simplified' guide to some appropriate statistical methods. The first section deals with the models and origins of hypotheses in human geography and the second is a tour of some of the statistical methods then used in geography. The last section on testing via analogues is concerned with simulation methods, and Haggett tantalizes the reader with an implicit suggestion that there might be an axiomatic approach to geography. He concludes with the observation: 'In the long run the quality of geography in this century will be judged less by its sophisticated techniques or its exhaustive detail, than by the strength of its logical reasoning'. A second edition of Locational Analysis appeared in 1977 – it was considerably revised and was almost double the length of the original. Andrew Cliff and Allan Frey were co-authors for this volume. It was also published as two separate paperbacks, entitled 'Locational Models' and 'Locational Methods' respectively (Haggett et al., 1977).

How the book was received

One of the first to comment publicly about Locational Analysis was Emrys Jones, a geographer at Queen's University, Belfast, in The Geographical Journal in 1966. Jones begins by warning readers of the some of the worst aspects of 1950's sociology: pseudo-mathematics, borrowed jargon, and quantification mania. Noting that recent developments in geography seemed to lack a coherent framework he praises Haggett for providing a framework for the new techniques and aims of geographical study. Jones summarizes the achievement of the book: 'Formerly disparate elements, like bits of a jigsaw puzzle, have clicked into place, and we are now confronted with a complete framework based on logical premises to enable us to describe location elements in an exciting way. This is brilliantly done' (Jones, 1966).

Peter Gould took a similar view in his review for the readers of the Geographical Review (Gould, 1967). His enthusiasm is palpable: 'here, at last, is a truly exciting work in human geography, splendidly organized, chock-full of ideas'; towards the end of his review he comments: 'taken as a whole this book not only gives a structure and coherence to the work of the last ten years … but it shows the continuity of such efforts with the best of the earlier traditions'. John Kolars (Kolars, 1967) presented a more sober review in Economic Geography in July 1967. He saw the 'value of this book is as a reference for the cognoscenti and as a guide for the catechumen'. Kolars describes the book as 'a synthesis of such proportions that it assumes a significance of its own'. Indeed, Haggett was to use the term synthesis in a later title of his own. Kolars also notes that for many sources, techniques and problems Haggett has drawn from

both physical and human geography, and goes on to suggest that by dropping the Human from its title allows it to express more clearly the spirit and purpose of geography as a discipline in the physical and social sciences.

The impact: what happened next?

The next few years saw an explosion of activity exploring the ideas so elegantly brought together in Locational Analysis. Several notable texts appeared, including Brian Berry and Duane Marble's (1968) Spatial Analysis, David Harvey's (1969) Explanation in Geography, and Ron Abler et al.'s (1971) Spatial Organization. New journals devoted to the spatial analysis were launched on both sides of the Atlantic, including Geographical Analysis (USA) and Environment and Planning (UK). A number of practitioners came together to create a forum for the exchange of ideas. The Quantitative Methods Study Group was incorporated by the Institute of British Geographers in 1969 and 'saw growth, expansion and widespread acceptance of quantitative methodology' (Gregory, 1983). The QMSG launched a series of undergraduate texts in 1975, the CATMOG (Concepts and Techniques in Modern Geography) series, edited by Peter Taylor. There were parallel developments in Europe – the first of what became a bi-annual colloquium was held in Strasbourg in 1978. The meeting brought together invited representatives from Britain, France, Austria, and Germany and has continued to stimulate much collaborative research. Moreover, the ideas from Geography started to diffuse into other disciplines, notably planning (which was in a growth phase during Britain's postwar reconstruction) as planners found themselves confronted with texts such as Models in Planning (Lee, 1973) and Urban and Regional Models in Geography and Planning (Wilson, 1974).

Other disciplines were raided. Location-allocation modelling (Rushton et al., 1973) had its origins in Operations Research, while geographer Alan Wilson introduced methods from statistical mechanics (Wilson, 1969) and pure mathematics (Wilson, 1981) into the arsenal of techniques. Ideas and approaches from statistics were borrowed too. The social sciences took on wholesale the ideas of classical statistical inference and used them as the engines of increasingly complex software packages such as SPSS, OSIRIS and SAS. Indeed SPSS proclaims itself as *the* Statistical Package for the Social Sciences. However, much had already taken place in sociology. For example, Social Area Analysis (Shevky and Bell, 1955) led to widescale experimentation with multivariate techniques such as factor analysis, discriminant analysis, canonical correlation analysis and cluster analysis. Gregory (1983) suggests with the benefit of hindsight we may regard some of the publications from the period as premature. However, the combination of enthusiasm and excitement is evident from the final words to an introductory text on Factor Analysis: 'grab some census data for your town, a factor analysis program and a large computer and run the data through the program … trying all the options' (Goddard and Kirby, 1976).

There were a few lone voices of caution – geographical data tended not to have the desirable properties that form the underpinning assumptions of classical statistics. Peter Gould was one – in his *statistics inferens* paper he sounded some warning bells (Gould, 1970). Others, having failed to change the world with large electronic computers and matrix inversion, tried radically different approaches. David Harvey argued that geographers had been worshipping at the wrong altar, and that there were richer opportunities elsewhere (Harvey, 1973). *Social Justice and the City* was a revelation for many as much as *Locational Analysis* had been a decade earlier. It is perhaps worthwhile considering an observation of the statistician Jan de Leeuw

on science: 'Science is, presumably, cumulative. This means that we all stand, to use Newton's beautiful phrase, "on the shoulders of giants"'. It also means, fortunately, that we stand on top of a lot of miscellaneous stuff put together by thousands of midgets. If we want to study a scientific problem we do this in the historical context, and we do not start from scratch. This is one of the peculiar things about the social sciences. They do seem to accumulate knowledge, there are very few giants, and every once in a while the midgets destroy the heaps' (de Leeuw, 1994). Indeed, Harvey was to describe the totality of the output from quantitative research as a 'hill of beans'. The debate between spatial scientists and social theorists continues today and there are those for whom spatial analysis is at best an irrelevance to the mainstream of the discipline (see Cloke *et al.*, 1991, amongst many).

A new technology appeared during the 1980s, that of Geographical Information Systems. The marriage of techniques from computational geometry, database systems, and computer science saw the emergence of a range of software packages for handling spatial data. Spatial analysis as practised with GIS deals with geometrical operations on spatial data. On both sides of the Atlantic government sponsored research initiatives were funded in the form of the ESRC's Regional Research Laboratory network and the National Center for Geographic Information and Analysis in the USA, accompanied by journals such as the *International Journal of Geographical Information Systems* and *Transactions in GIS*. The rise of GIS has been paralleled by a return to the problems of statistical modelling with spatial data. Practitioners have not only included geographers but also statisticians.

Today, spatial data handling software is ubiquitous, and appears in a wide variety of forms. Dozens of airlines have on-line ticket ordering systems which will provide the traveller with the quickest or cheapest ticket option. Planning rail or road journeys via the Internet uses the same underlying models and methods. Satellite navigation systems are standard items in high street electronics stores. Geometry and geography make interesting bedfellows and computational geometry puts the progeny in the hands of the public. The challenging problems in locational analysis remain challenging – and there is no shortage of researchers willing to tackle them.

Conclusion

Haggett's book stands as an early intervention in geography's attempt to restyle itself as a spatial science, and remains an obligatory point of passage for many geographers working on spatial data analysis, GIS and geocomputation. Trevor Barnes' summation of the book is thus as follows:

> On the surface, its limpid prose, elegant diagrams, and mathematical equations are not the sort of thing to get people riled up to change the world. But from recent interviews I've carried out in connection with a project to write about the history of the quantitative revolution in geography during the 1950s and 1960s, and economic geography in particular, it is clear that Haggett's book had exactly that effect. Moreover, its influence was not just confined to within the academy, persuading people merely to write a different type of academic paper. (Barnes, 2002: 10)

The material impacts of Haggett's book on his contemporaries and subsequent generations were then, in many cases, massive. As such, there is little doubt that Locational Analysis is one of human geography's Key Texts – one whose legacies continue to unfold.

Secondary sources and references

Alber, R., Adams, J.S. and Gould, P. (1971) *Spatial Organisation: The Geographer's View of the World*, Englewood Cliffs, NJ: Prentice-Hall.

Barnes, T.J. (2002), 'Critical notes on economic geography from an ageing radical', *Acme* 1 (1): 8–20. http://www.acme-journal.org/vol1/barnes.pdf

Barnes, T.J. (2003) 'The place of locational analysis: a selective and interpretive history', *Progress in Human Geography* 27 (1): 69–85.

Berry, B.J.L. and Marble, D. (1968) *Spatial Analysis: A Reader in Statistical Geography*. Englewood Cliffs: Prentice-Hall.

Bunge, W. (1962) 'Theoretical geography', Lund Studies in Geography, series C, *General and Mathematical Geography* 1.

Chorley, R.J. and Haggett, P. (1965) 'Trend-surface mapping in geographical research', *Transactions of the Institute of British Geographers* 37: 47–67.

Cloke, P., Philo, C. and Sadler, D. (1991) *Approaching Human Geography: An Introduction to Contemporary Theoretical Debates*. London: Paul Chapman.

De Leeuw, J. (1994) Statistics and the Sciences. Unpublished manuscript, UCLA Statistics Program.

Evenden, J.Y. (1969) 'Review: Innovation diffusion as a spatial process' by Törsten Hägerstrand, *Social Forces* 47 (3): 356–357.

Gehlke, C.E. and Biehl, K. (1934) 'Certain effects of grouping upon the size of the correlation coefficient in census tract material', *Journal of the American Statistical Association* 29, Supplement: 169–170.

Goddard, J.B. and Kirby, A. (1976) *An Introduction to Factor Analysis*. Norwich: Geo Abstracts Ltd.

Gould, P. (1967) 'Review: Locational analysis in human geography', *Geographical Review* 57 (2): 292–294.

Gould, P. (1970) 'Is statistix inferens the geographical name for a wild goose', *Economic Geography* 46, Supplement: 439–448.

Gregory, S. (1965) 'Regression techniques in geography', *Transactions of the Institute of British Geographers* 37: ix–xi.

Gregory, S. (1983) 'Quantitative geography: the British experience and the role of the Institute', *Transactions of the Institute of British Geographers, New Series* 8: 80–89.

Hägerstrand, T. (1967) *Innovation Diffusion as a Spatial Process*. Chicago: University of Chicago Press.

Haggett, P. (1965) *Locational Analysis in Human Geography*. London: Edward Arnold.

Haggett, P., Cliff, A.D. and Frey, A. (1977) *Locational Analysis in Human Geography* (2nd edn). New York: Wiley.

Harvey, D.W. (1969) *Explanation in Geography*. London: Edward Arnold.

Harvey, D.W. (1973) *Social Justice and the City*. London: Edward Arnold.

Jones, E. (1966) 'Review: Location in geography: a statistical approach', *The Geographical Journal* 132 (2): 267–268.

Kolars, J. (1967) 'Review: Locational analysis in human geography', *Economic Geography* 43 (3): 276–277.

Lee, C. (1973) *Models in Planning: An Introduction to the use of Quantitative Models in Planning*. Oxford: Pergamon.

Neprash, J.A. (1934) 'Some problems in the correlation of spatially distributed variables', *Journal of the American Statistical Association* 29, Supplement: 167–168.

Rushton, G., Goodchild, M.F. and Ostresh, L.M. (1973) *Computer Programs for Location-Allocation Problems*. Monograph No. 6, Department of Geography, University of Iowa.

Shevky, E. and Bell, W. (1955) *Social Area Analysis: Theory, Illustrative Application and Computational Procedures*. Stanford: Stanford University Press.

Wilson, A.G. (1969) *Entropy in Urban and Regional Modelling*. London: Centre for Environmental Studies.

Wilson, A.G. (1974) *Urban and Regional Models in Geography and Planning*. London: Wiley.

Wilson, A.G. (1981) *Catastrophe Theory and Bifurcation: Applications to Urban and Regional Systems*. London: Croom Helm.

Wooldridge, S.W. and East, W.G. (1951) *The Spirit and Purpose of Geography*. London: Hutchinson.

4 EXPLANATION IN GEOGRAPHY (1969): DAVID HARVEY

Ron Johnston

By our theories you shall know us. (Harvey, 1969: 486)

Introduction

Although the discipline of geography has always been characterized by flux, the 1960s is recognized as a particular turbulent decade: by its end, the discipline incorporated practices very different from those deployed 10 years earlier. Particularly important here was the discontent with disciplinary practices which some geographers felt after having served with scholars from other disciplines in the American Office of Strategic Services during World War Two (Barnes and Farrish, 2006). Subsequently in the mid-1950s a number of US geographers – notably a cohesive group of faculty and graduate students at the University of Washington, Seattle – began to promote a very different vision of geography (bridging both physical and human arms of the discipline) based on the 'scientific methods' deployed by physicists and, in the social sciences, economists. By the early 1960s, this group was rapidly attracting adherents in a number of major US graduate schools, and in 1963 one observer-participant claimed that the 'revolution' promulgated from Seattle was successfully over.

Almost contemporaneously, a similar 'revolution' was taking shape in the UK, based at the University of Cambridge. Two recently appointed faculty members – Dick Chorley and Peter Haggett – were attracted to the 'scientific model' (which Chorley had encountered and

adopted while a geology graduate student in the US), and began to teach related material, mainly statistical methods, in their introductory practical classes for geography undergraduates. One of the demonstrators in the first year of that class (in 1960) was a postgraduate student working in historical geography – David Harvey (Harvey's 1962, PhD thesis was on the historical geography of the Kentish hop industry – see Harvey, 1963 – and this stimulated his subsequent interest in the processes involved in evolving spatial patterns: Harvey, 1967).

Although the shift in geographical practice these two groups were advancing is often referred to as the discipline's 'quantitative revolution' (on which, see Johnston and Sidaway, 2004), it involved much more than just applying mathematical and statistical techniques to geographical data. It was fundamentally a 'theoretical revolution', which changed the entire mind-set of how research was to be undertaken and new knowledge presented. Words such as 'theory', 'model', 'hypothesis' and 'law' become common in the geographical lexicon, as researchers strove to produce knowledge that was *cumulative* (in the sense that it built on earlier research not only to better explain the world but also to modify or improve it).

When a discipline is experiencing such major change, new teaching materials are needed to introduce students to (and justify) 'revolutionary' practices. Introductory textbooks usually lag behind such changes given that publishers have to be convinced that there is a viable market for 'revolutionary tracts'!

This was certainly the case with the 1960s 'new geography'. The first books clearly enunciating the 'scientific approach' appeared in 1965 (Haggett's *Locational Analysis in Human Geography* – this was very much a publisher's gamble on the shape of the discipline's future (see Charlton, Chapter 3 this volume) – and Chorley and Haggett's edited volume *Frontiers in Geographical Teaching*; Bunge's (1962) monograph on *Theoretical Geography* had only a limited circulation, although a revised edition in 1966 attracted wider attention – see Goodchild, Chapter 2 this volume). A spate of texts oriented towards the 'revolution' only emerged some five years later. This included David Harvey's *Explanation in Geography*, the product of almost a decade teaching undergraduates at the University of Bristol about the 'new' scientific basis to geographical work, building on discussions with colleagues in Sweden and the US (where he spent considerable time during the decade). As he puts it in the Preface, writing the book was part of his learning experience as he developed teaching materials – he 'wrote this book in order to educate myself. I sought to publish it because I feel sure there are many geographers, both young and old, who are in a similar state of ignorance to that which I was in before I commenced to write' (Harvey, 1969: v).

The book and its argument

Harvey's book – like that of the other 'revolutionaries' he joined – reflected a deep dissatisfaction with geographical practices experienced when he was an undergraduate and in his formative years as a researcher. For Harvey, quantification was necessary but far from sufficient: measurement was a requisite tool, but much more important was for human and physical geographers to deploy the 'fantastic power' of the scientific model. Hence he explored – and wrote a book (he termed it an 'interim report') about – 'the ways in which geographical understanding and knowledge can be acquired and the standards of rational argument and inference that are necessary to ensure that the process is reasonable' (Harvey, 1969: viii). This exploration took him into a literature previously almost entirely ignored by geographers, so that although he included plentiful references to a small number of geographers (especially those instrumental in fomenting the 'revolution') his 'Author index' indicates how heavily he drew on philosophers of science – Ackoff, Braithwaite, Carnap, Churchman, Hempel, Kuhn, and Nagel; Popper is also cited, though to a lesser extent (and his ideas regarding falsification are very summarily dismissed) and even Einstein and Russell (although neither Ayer nor Wittgenstein) are mentioned. In addition, a significant number of mathematicians and statisticians such as Anscombe, Blalock, Fishburn, Fisher, Krumbein, Sneath, and Sokal are also cited.

From the outset, Harvey's book focuses explicitly on explanation. After a brief introductory chapter setting out his main concern as methodology rather than philosophies – i.e. on how geographers should produce explanations – *Explanation in Geography* moved to a discussion of what explanation means. Harvey defined it as 'making an unexpected outcome an expected outcome, of making a curious event seem natural or normal' (Harvey, 1969: 13) because it can be shown to be generated by similar processes and in similar conditions as previous events of the same type. Harvey's concern was with 'rational explanation', statements verifiable by others because the procedures involved in their production can be repeated and/or are open to scrutiny. It is a 'formal procedure … [at the] hard inner core of methodology' (Harvey, 1969: 23).

Following that brief framework-setting introduction (of only 26 pages), the remainder of the book comprised five main sections dealing with: explanation in geography; theories, laws and models; languages; models for description in geography; and models for explanation in geography. At the outset he contrasted (in an oft-reprinted diagram) the inductive or 'Baconian' path to explanation with his preferred deductive route, which proceeds through the establishment of a model representing the researcher's image of the world, the derivation of hypotheses regarding some aspect of that image, testing the hypothesis's validity, and the formulation of theories and laws synthesizing the knowledge gained – from which revised models can be derived.

Four terms/concepts stood out in the lexicon Harvey associated with this route to explanation. *Hypotheses* were presented as logically consistent 'controlled speculations'. They can never be tested absolutely – conclusions are always provisional – but Harvey indicated that they guide the production of scientific knowledge in a rigorous (and hence replicable) way. The key outcomes of testing hypotheses are laws and theories. *Laws* are sometimes presented as universal truths – statements whose validity is constrained by neither time nor space. It is never possible to reach such conclusions, but scientists act as if they can, using their findings as statements of the current 'conventional wisdom' encapsulating what we already know and providing the foundations for further scientific exploration. They are rigorously produced but ultimately provisional conclusions representing the contemporary state of knowledge. For geography (especially human geography, although much physical geography at the time – such as the Davisian model of landscape evolution – similarly lacked rigorous underpinnings), they contrasted sharply with the 'explanatory sketches' traditionally offered as accounts of 'observed reality'. The production of laws in geography, according to Harvey, involved searching for 'hidden order within chaos'. Because the results of such searches are likely to be provisional/tentative, Harvey suggested that geographers might prefer the concept of 'law-like statements', general claims that are both 'reasonable with respect to experience and consistent with respect to each other' – a coherent body of knowledge corresponding with the observed 'reality'.

Theories are systems of linked statements about defined subject matter. These may be entirely closed systems – as with Euclidean geometry; they may be sets of deductive statements derived from accepted axioms; or they may be less-formally stated 'sketches'. They are not just speculative ideas – as sometimes implied in vernacular uses of the term: anyone can fabricate such a 'system of apparent wisdom in the folly of hypothetical delusion' (Harvey, 1969: 97: here, Harvey is quoting James Hutton via Chorley). Scientific theories take 'such speculations and transform ... them from badly understood and uncomfortable intrusions upon our powers of "pure" objective description into highly articulate systems of statements of enormous explanatory power' (Harvey, 1969: 87–88).

Various types of theory extend along a continuum from highly formalized, internally-closed sets of statements (as with many forms of mathematics and logic), through sets of statements which are only partial (either because their primitive terms, or axioms – the assumptions on which they are built – are incomplete or because the deductions from those foundations are not fully elaborated), to what Harvey terms 'non-formal theories ... statements made with theoretical intention, but for which no theoretical language has been developed' (Harvey, 1969: 98). The last type 'scarcely conform in any respect to the standards of scientific theory' (Harvey, 1969: 130), and are characteristic of previous 'theoretical' work within geography. Geographers had to move forward, Harvey argued, either by deriving theories from axiomatic statements – more likely to be feasible in physical geography, which can deduce, for example, landscape-producing processes and the likely resultant forms from physical laws – or, in human geography, by generating assumptions about human behaviour from which statements about spatial patterns can be deduced.

Whatever the theory's origin, establishing its empirical status involves moving to the final key word in the new lexicon Harvey employed – the *model.* This has a diversity of meanings in both popular and scientific language; for Harvey's 'new geography' a model was a representation of a theory – i.e. an outcome of a series of law-like statements. Such representations became the source for hypotheses, leading to tests of a theory's empirical validity.

An example of a (human) geographical theory illustrating these fundamental concepts is *central place theory*: an idea about the spatial organization of settlement hierarchies that underpinned a great deal of geographical work in the 1950s–1960s. Harvey showed that it was derived from a set of fundamental economic postulates (assumed laws) about consumer and provider behaviour (i.e. profit-maximization for providers and minimization of travel costs for consumers) and the nature of the goods/services being supplied/demanded. These are linked in a single theory from which it is possible to deduce the spatial arrangement of service centres. Models could be derived showing the expected morphology of that spatial arrangement in different contexts, which had common features such as the hexagonal arrangement of centres in nesting hierarchies. Specific hypotheses could then be tested in particular empirical situations.

For Harvey, the most important of these concepts was theory: without theories 'the explanation and cognitive description of geographic events is inconceivable' (Harvey, 1969: 169). But how could such theories be expressed? The language to be deployed was that of mathematics – 'the language of science' – within which he concentrated on two sub-fields: *geometry* as the language of spatial form (geography being defined as the study of 'objects and events in space' – Harvey, 1969: 191); and *probability* as the language of chance, necessarily used because 'the world is governed by immutable chance processes' (Harvey, 1969: 260) so that precise prediction is rarely possible, especially so given the extent of our ignorance about those processes. A scientific geography would not be a deterministic geography, therefore, but rather comprise probabilistic statements of likely explanations (of the 'hidden order within chaos') – hence Harvey's advocacy of statistics as crucial in evaluating hypotheses.

Given the key components of the scientific method and its language, as applied to geography (both human and physical; Harvey saw no difference between the two in their methodological structure), Harvey thus dedicated two sections of the book to modelling in geography – descriptive and explanatory. These are, in effect, chapters about methods: the first deals with measurement and how one portrays the world – how information is collected, classified and displayed; the second with procedures for testing hypotheses of cause and effect.

In the concluding chapter, Harvey summarized 480 pages of detailed material as 'some rough and ready guidelines for the conduct of empirical research in geography', presenting the tools that might be used when we 'have to pin down our speculations, separate fact from fancy, science from science fiction' (Harvey, 1969: 481). That is done scientifically by producing 'an adequate corpus of geographic theory' – coherent statements about aspects of the world, validated by geographers' adoption of the protocols of scientific method. For Harvey geography in the 1960s lacked such a clear identity and sense of direction, hence the clarion call in the final pages of his text:

> Without theory we cannot hope for controlled, consistent and rational explanation of events. Without theory we can scarcely claim to know our own identity. ... theory construction on a broad and imaginative scale must be our first priority in the coming decade. ... Perhaps the slogan we should pin upon our study walls for the 1970s ought to read: 'By our theories you shall know us'. (Harvey, 1969: 486)

The book and its consequences

Explanation is a long and detailed book. It is not difficult to read once the basic concepts are appreciated, but it is not presented as a textbook – or at least not as such books are now presented (i.e. there are suggested readings at the end of each chapter, but no boxes or other devices designed to focus on key ideas and/or exemplars). Harvey never intended to write such a textbook: as he expressed it, the book was 'written for anyone who cares to read it and for anyone who cares to use it in whatever way or ways they find congenial or useful. ... The sooner we stop writing for 'an' audience ... or for 'the beginning graduate student' the better off we will be' (Harvey, 1971: 323). His goal was nevertheless that of other contemporary textbook writers: to introduce the methods geographers (should) deploy to produce knowledge, beginning with the key methodological protocols and then setting out detailed procedures. It is thus similar to other contemporary volumes, notably Abler *et al.* (1971), which began with discussions of 'scientific method' and research procedures but then, in a much more 'student-friendly' way, illustrated these with detailed (human) geographical examples; Haggett (1965) put procedures after examples – they became separate books in the second edition – but said very little about 'scientific method'. Earlier books, such as Gregory (1963), were entirely about procedures (interestingly, Gregory's book is not referenced in *Explanation*).

Explanation's accessibility – at least to Harvey's peers – was testified by reviews at the time. Among the main reviews are those by Douglas Amedeo (1971), for whom it was clear that 'geographers need this kind of book, and it would be a profound disservice to the field to provoke them into ignoring it'; Stan Gregory (1970) who called it 'an important book, which is certain to have a considerable influence upon the development of geography over the next decade'; and Julian Wolpert (1971) who termed it 'enormously instructive', suggesting Harvey 'has contributed to the discipline a rigorously concise overview of the accumulated "science" which converts current research and teaching efforts ... [providing] geographically-relevant

access to the mainstream of the philosophy, methodology, and language of science'. These reviews depicted the book as important, especially as it was the first full-length statement of what might be involved in geographers adopting the methodology (and underlying philosophy) of the natural sciences – what geographers later referred to with increasing frequency as *positivism* (a term not in *Explanation*'s index!).

But *Explanation*'s impact was probably less than its originality and depth deserved, given that it was the first full statement of what a 'scientific geography' should look like. For example, *Explanation* is unlikely to have been widely used as a textbook – certainly not for undergraduates – for three main reasons. The first is its density and presentation. Many in the early 1970s will have been directed to it as supplementary reading for the increasingly popular courses on the history and philosophy of geography – but it was too detailed to be deployed as the main text for a course on geographical methods (Harvey's course at Bristol was a rarity within undergraduate programmes at the time and the book appeared more than a decade before taught postgraduate courses were instituted in many British geography departments: it may have attracted more attention in US graduate schools, but competed with more 'accessible' books such as Abler *et al.*, 1971). Some undergraduates will have been invited to read particular parts as the basis for tutorial/seminar discussions; others may have either discovered it serendipitously on a library/bookshop shelf or been directed towards it, and become absorbed by its arguments. But for most students (and also academics intrigued by and attracted to the changes in their discipline) its main role was as a reference text, something to be used when seeking detailed material about, for example, the key concepts of theory, law, hypothesis and model. Notably, it offered few examples nor the case studies that are so often the key to student appreciation of a set of ideas, and certainly its chapters on procedures did not take the 'how-to' form typical of introductory text-books (for example, Amedeo and Golledge, 1975, was a more 'user-friendly' introduction to the protocols of the scientfic method).

Secondly, although original in its depth and catholicity *Explanation* was not entirely novel. 'The shock of the new' had hit geographers some years before with the publication of books such as Chorley and Haggett (1965), Haggett (1965) and Chorley and Haggett (1967). These were much more instrumental in bringing the fomenting revolution in geographical practices to academic geographers' attention (especially in the UK). For those converted, however, *Explanation* provided the detailed exposition needed to make them fully aware of the complexity of the practices they were planning to adopt. But, as Harvey made clear, the book was not about philosophy but rather methods within a particular philosophy, and although he clearly addressed many philosophical issues, it was left to later authors (notably Gregory, 1978) to explore issues of epistemology and ontology more fully, linking geographers to a much wider set of philosophical debates.

Thirdly, by the time *Explanation* appeared the practices that it advanced were being strongly contested – not least by David Harvey himself! He hints at this in his Preface:

> Compared to my situation five years ago I now feel much more learned and wise, but relative to what I still have to learn I feel more ignorant than ever. Indeed, since completing this manuscript in June 1968 I have changed several opinions and I can already identify errors and shortcomings in the analysis. (Harvey, 1969: viii)

Thus, when he responded to by far the longest review/critique to appear – by Stephen Gale (1971), who thought it ambitious and stimulating, but deficient as both textbook and reference volume – Harvey had moved away, not from science *per se* but from that particular scientific philosophy and form of scientific method. As detailed in a number of essays in the early 1970s (reprinted in Harvey, 1973), he turned to Marxism as his source of theoretical inspiration, claiming the scientific method he had previously advanced was ideologically infused in that it sustained the political status quo and, however sophisticated its descriptions of the world, was unable to appreciate the underlying processes which produced them (see Harvey, 1974). Harvey certainly did not abandon theory: rather, he moved his theoretical stance, adopting a new set of protocols and procedures. As Peet (1998) put it, with *Explanation* Harvey showed geographers they needed theory, but almost immediately realized it was the wrong theory: his clarion call accordingly remained intact – but the theories did not.

The book's legacy

Although there was a clear 'revolution' in David Harvey's own approach to human geography almost immediately after *Explanation* was published, quantitative work remains a strong strand within the discipline – increasingly sophisticated in both its methods and in the computational, statistical and GIS technologies on which it now depends. But much of that work has increasingly distanced itself from its positivist foundations – particularly with regard to the search for law-like statements. As discussed in recent essays (such as Fotheringham, 2006), most quantitative work involves the rigorous interrogation of large, spatially structured data sets with the goal of finding order within a highly complex world (but never implying that such order is fixed for eternity). Such work seeks to accrue 'sufficient evidence on which to base a judgement about reality that most reasonable people would find acceptable' (Fotheringham, 2006: 241), in contexts where understanding calls for the deployment of large, aggregate data sets. For such enterprises, much of the detail in *Explanation* is irrelevant given that many of the methods it promotes have been superseded. Indeed, while *Explanation* was one of the foundation stones in the creation of a 'new theoretical geography', the superstructure has since been substantially reconstructed in subsequent decades.

In retrospect, it is clear that *Explanation* was written on the cusp of a major change in Harvey's own work in the early 1970s (as Harvey, 2002, 2006, documents; see also Castree, Chapter 8, this volume). In spite of, and perhaps because of, that he remains one of geography's most influential scholars. He has also – uniquely – been the subject of two critical evaluations. The first (Paterson, 1985) covered the *Explanation* years and the first decade of Harvey's Marxist explorations, but gave less than one-third of its space to the former and, while documenting the major discontinuity in his thought, gave little attention to what might be identified as a major continuity – 'the importance of general theory'. (This misrepresents Harvey for whom theory is the continuity, not 'general theory', whatever that might be.) The second (Castree and Gregory, 2006) appeared some 35 years after the switch in Harvey's theoretical orientation and, given his productivity and the seminal nature of much that he has published since, not surprisingly pays relatively little attention to the first decade or so or Harvey's career. Trevor Barnes puts that 'first life' into its social context and stresses its continuities with Harvey's post-1970 output – commitments to the discipline and practice of geography, to politics (i.e. the application of geographical knowledge), and 'perhaps most germane…, to theory' (Castree and Gregory, 2006: 42); in the same volume, Derek Gregory unpacks the lacunae in *Explanation* – the black-box system diagrams that expressed an ignorance of process and the sterility of mathematical language, while Eric Sheppard identifies several more continuities, notably a concern for space and time – and space–time. But it is Harvey the Marxist who gets the bulk of the attention (as in Castree, 2004).

Conclusion

To paraphrase a statement made in another context, *Explanation* is undoubtedly now 'more revered than read' – and may always have been so. As Castree (2004: 181) puts it, *Explanation* 'gave Harvey's generation of geographers a heavyweight justification and manifesto for their project', aligned 'the discipline with the so-called "real" sciences like physics and, for some geographers, boosted the discipline's self-image' – although many human geographers sought status within the social sciences rather than suffering from 'physics envy'. But Harvey abandoned his generation – or many of them – for an alternative project, to which a new generation of converts was attracted. Indeed, according to his autobiographical essay, in some ways he abandoned the first project long before he completed it – having a 'lust to wander and diverge, to challenge authority, to get off the beaten path of knowledge into something different, to explore the wild recesses of the imagination as well as of the world' (Harvey, 2002: 167). He did finish it, however, but responded to Stephen Gale's (1971) review by saying that he was at a disadvantage because Gale had read the book and 'I have never read it. What is more, I have no intention of doing so now'. *Explanation* was behind him, but remains a permanent and potent reminder of a crucial time in geography's turbulent recent history; as such, it should be read not just as a pioneering and influential exploration of 'scientific method' and its philosophical underpinnings but also one of the first substantive geographical engagements with social *science*.

Secondary sources and references

Abler, R., Adams, J. and Gould, P. (1971) *Spatial Organization: The Geographer's View of the World*. Englewood Cliffs, NJ: Prentice-Hall.

Amedeo, D. (1971) 'Review of Explanation in Geography by David Harvey', *Geographical Review* 61: 147–149.

Amedeo, D. and Golledge, R.G. (1975) *An Introduction to Scientific Reasoning in Geography*. New York: John Wiley.

Barnes, T.J. and Farrish, M. (2006) 'Between regions: science, militarism, and American geography from World War to Cold War', *Annals of the Association of American Geographers* 96: 807–826.

Bunge, W. (1962) *Theoretical Geography* (second edition 1966). Lund: C.W.K. Gleerup.

Castree, N. (2004) 'David Harvey', in P. Hubbard, R. Kitchin and G. Valentine (eds) *Key Thinkers on Space and Place*. London: Sage, 181–188.

Castree, N. and Gregory, D. (eds) (2006) *David Harvey: a Critical Reader*. Oxford: Blackwell Publishing.

Chorley, R.J. and Haggett, P. (eds) (1965) *Frontiers in Geographical Teaching*. London: Methuen.

Chorley, R.J. and Haggett, P. (eds) (1967) *Models in Geography*. London: Methuen.

Fotheringham, A.S. (2006) 'Quantification, evidence and positivism', in S. Aitken and G. Valentine (eds) *Approaches to Human Geography*. London: Sage, 237–250.

Gale, S. (1971) 'On the heterodoxy of explanation: a review of David Harvey's Explanation in Geography', *Geographical Analysis* 3: 285–322.

Gregory, D. (1978) *Ideology, Science and Human Geography*. London: Hutchinson.

Gregory, S. (1963) *Statistical Methods and the Geographer*. London: Longman.

Gregory. S. (1970) 'Review of Explanation in Geography by David Harvey', *The Geographical Journal* 136: 303.

Haggett, P. (1965) *Locational Analysis in Human Geography*. London: Edward Arnold.

Harvey, D. (1963) *Explanation in Geography*. Arnold: London.

Harvey, D. (1963) 'Locational change in the Kentish hop industry and the analysis of land use patterns', *Transactions, Institute of British Geographers* 323: 123–140.

Harvey, D. (1967) 'Models of the evolution of spatial patterns in human geography', in R.J. Chorley and P. Haggett (eds) *Models in Geography*. London: Methuen, 549–608.

Harvey, D. (1969) *Explanation in Geography*. Arnold: London.

Harvey, D. (1971) 'On obfuscation in geography: a comment of Gale's heterodoxy', *Geographical Analysis* 3: 323–330.

Harvey, D. (1973) *Social Justice and the City*. London: Edward Arnold.

Harvey, D. (1974) 'What kind of geography for what kind of public policy', *Transactions, Institute of British Geographers* 63: 18–24.

Harvey, D. (2002) 'Memories and desires', in P.R. Gould and F.R. Pitts (eds) *Geographical Voices: Fourteen Autobiographical Essays*. Syracuse: Syracuse University Press, 149–188.

Harvey, D. (2006) 'Memories and desires', in S. Aitken and G. Valentine, (eds) *Approaches to Human Geography*. London: Sage, 184–190.

Johnston, R.J. and Sidaway, J.D. (2004) *Geography and Geographers: Anglo-American Human Geography since 1945 (sixth edition)*. London: Arnold.

Paterson, J.L. (1985) *David Harvey's Geography*. London: Croom Helm.

Peet, R. (1998) *Modern Geographical Thought*. Oxford: Blackwell.

Wolpert, J. (1971) 'Review of Explanation in Geography by David Harvey', *Annals of the Association of American Geographers* 61: 180–181.

5 CONFLICT, POWER AND POLITICS IN THE CITY (1973): KEVIN COX

Andy Wood

The American city is in a state of crisis. The melting pot of yesterday has become a Pandora's box of troubles – flight to the suburbs, ghetto poverty, racial conflict, inadequacies in public provision are fodder for contemporary urban politics. Conflict has become endemic in the metropolitan areas: between the 'turfs' of social groups, between suburbs and central city, and between neighborhoods and the city itself. This book is concerned with the geography of these conflicts. (Cox, 1973: 1)

Introduction

Kevin Cox's third book *Conflict, Power and Politics in the City: A Geographic View*, had the misfortune to be published in 1973 – the same year as David Harvey's *Social Justice and the City*. Citation counts indicate the tremendous significance of the latter work but if Harvey's overshadowed Cox's book the latter nevertheless remains one of the key texts from the early 1970s that put urban geography at the forefront of disciplinary trends (with Ley, 1974 and Ward, 1971 also fitting that bill). While human geography has, in the main, moved on from the intellectual ideas that underpin *Conflict, Power and Politics in the City* the book stands as an important transitional document that has remarkable contemporary resonance even though the theoretical tide has turned.

The combination of Preface and Conclusion indicates that *Conflict, Power and Politics in the City* was written to meet three important goals. The first was to advance a rigorous, academic account that would help to explain the urban problems that were increasingly apparent in US cities such as Cleveland, Los Angeles, Newark and New York in the late 1960s and early 1970s. In fact rioting and violent political protest have a long history associated with the city but the urban conflagrations of the late 1960s threatened to thoroughly disrupt the idea that the US was a nation of growing wealth and prosperity for all. It is indeed notable that *Conflict, Power and Politics in the City* was published in McGraw-Hill's *Problems Series in Geography*. Earlier volumes had addressed atmospheric pollution (Bach, 1972), poverty (Morrill and Wohlenberg, 1971) and social well-being (Smith, 1973). The series gave Cox the opportunity to develop an academic text that was explicitly focused on contemporary urban problems while avoiding the more particular focus of Harold Rose's 1971 book *The Black Ghetto* in the same series.

A second goal of Cox's book was, as he suggests (Cox, 1973: 1), to publicize the policy implications of urban analysis, with the final chapter devoted to a lengthy discussion of what he saw as the necessary governmental actions required to address the conflicts examined in his book. The third objective was one of developing a specifically *geographical* account of urban conflicts and the processes that underpin them. In this respect Cox's immediate academic environment seems to have been significant. Kevin Cox had grown up near Warwick in England and had studied Geography at Cambridge. He then traveled to the University of Illinois where he completed a Master's degree and his doctorate. In 1965 the editor of the McGraw-Hill series – Ned Taaffe – had recruited Cox to Ohio State as part of a strategy of hiring what were to become key figures in the discipline, including Larry Brown, Emilio Casetti, George Demko, Howard Gauthier, Reginald Golledge, Leslie King and John Rayner (Barnes, 2004). The intellectual coherence and direction to the department was provided by an uncompromising focus on quantitative geography and the prosecution of the spatial science tradition. Trevor Barnes has written that during the 1960s 'perhaps more than at any other place, OSU (Ohio State University) self-consciously remade itself into a site of quantitative geography' (Barnes, 2004: 582). Without wanting to claim that these biographical details determined Cox's position it is clear that this immediate environment had significant influence on the nature and direction of Cox's venture.

Conflict, Power and Politics in the City: the book and its arguments

The book, which runs to a trim 133 pages, is organized into five chapters. Chapter One introduces the basic concepts that provide the foundation for Cox's framework. Chapter Two, the shortest, addresses the territorial organization of metropolitan areas and establishes the context for the two meatier chapters that follow. The first of these (Chapter Three) examines the relationship between 'metropolitan fragmentation and urban conflict,' focusing largely on fiscal or monetary disparities between central city and suburb. Chapter Four then switches geographical scales to examine conflict within

the city. The concluding chapter, as noted above, addresses the policy implications of the foregoing analysis.

Chapter One establishes the theoretical basis for the book and it is worth spending some time outlining the basic argument. Following the well-worn tradition of the spatial science that defined his department, Cox starts out with a set of simplifying assumptions about the nature of economic and political life. Cox initially asks the reader to assume a purely private economy comprising individual decision-making units in the form of households, firms and other organizations, each of which has a 'utility-function' that specifies preferences for different goods and services or 'commodity bundles.' Each unit allocates resources in trying to maximize its utility by securing its preferred mix of goods and services. Although the terms may seem quite peculiar now, the basic argument is a very familiar one for the time.

Where Cox departs from the conventional view, however, is through the complication that individual utilities are 'not independent of the resource allocations of others' (Cox, 1973: 2). The thoroughly *social* nature of urban life proves to be a key theme in Cox's account. The fact that an individual's utility is influenced by the resource allocation of others generates what Cox terms 'externality' or 'spillover effects.' Externality effects can be either positive or negative depending upon the nature of what they provide. Improvements that 'beautify' or improve a neighborhood – such as parks or 'good' schools – are seen as generating positive externalities to property owners within that neighborhood. In contrast, the introduction of 'noxious' sources or facilities such as sewage farms (water-treatment plants) or homeless shelters generate externalities that are likely to be perceived as negative by those affected by them. Given the fundamental assumption that individual decision-making units will always maximize their own utility, Cox argues that, for society as a whole, negative externalities or indirect costs tend to be 'overproduced' while positive externalities or indirect benefits tend to be 'underproduced.' Accordingly, externality effects 'detract from overall social welfare and pose serious problems for society as a whole as well as for the individuals that make up that society' (Cox, 1973: 3).

While externality effects are pervasive, Cox insists they are not spatially random. Instead typical externalities are seen as explicitly geographical in that the intensity of externality effects varies with location relative to the source of the externality. For example, those that live adjacent to a park are seen to gain greater benefit than those at some distance from it. Likewise, those next to a refuse plant or power station are more likely to be affected by pollution and noise than those further away. Furthermore, externality effects in the city are magnified by the relative proximity of decision-makers to one another and thus to the externalities that result from their decision-making. In short, *externality effects* – whether negative or positive – are hard to avoid in the urban context. Cox attributes much of the variation in local environmental quality within the city to externality effects with the basis of locational conflict seen as rooted in the conflict between the attempt to maximize individual utility and the externality generating effects of other proximate decision-makers.

Employing a simple symmetry, Cox argues that if locational conflicts are the product of externality effects then the resolution of those conflicts must involve coordinating the resource allocations that produce these outcomes. To that end there are two basic coordinating mechanisms or strategies. The first involves the private coordination of activities in which individual decision-makers seek to mitigate the impact of negative externalities. The second requires public intervention through the centralized, collective decision-making apparatus of government. In the case of private coordination Cox identifies two

alternative courses of action available to decision-makers. The first – which is essentially an 'exit' option – involves relocation. The second is to address immediate externality effects by bargaining *in situ* with those that produce them. Cox highlights a range of hypothetical outcomes resulting from the two private strategies and the conceptual model provides a clear and robust basis for examining alternative courses of action. Cox argues that solutions involving relocation predominate or, in simple terms, residents generally prefer to exit rather than to bargain. Rather than producing a locational solution maximizing the utility of all, the result is the very familiar metropolitan pattern in which the city is organized or segregated into discrete neighborhoods, each of which is relatively homogenous in terms of income and race. Cox (1973: 9) argues that such patterns of stratification reflect the close relationship between income, race and the type and nature of externality effects.

The product of such strategies is clearly not optimal given the basic contradiction between individual utility maximization and the general welfare of society. For Cox the means for addressing this tension is found not in the individual coordinating efforts of private individuals but rather in the coercive and binding, collective decision-making power of government. Governments are seen as particularly important in providing 'public goods' – goods that, following Hirschman (1970), are nonexclusive and equally available. These public goods, which would not be produced by private means, play an important coordinating role with respect to externalities. Public health and environmental legislation, for example, is designed to limit some of the most egregious negative externalities, while other forms of regulation, such as zoning, seek to eliminate externalities resulting from the juxtaposition of incompatible land uses.

Whereas private forms of coordination generate a *de facto* pattern of spatially discrete neighborhoods, Cox argues that public coordination gives rise to a *de jure* pattern of jurisdictional spaces. Again coordination is seen to mitigate but not eliminate negative externality effects. The locational conflicts that result are neatly divided into those *between* jurisdictions and those *within* jurisdictions. Exclusionary zoning on the part of US suburbs provides a very clear example of the way in which the actions of one governmental unit can have detrimental impacts on others, not least those subsequently excluded from suburban housing, such as the low-income populations of the central city. A second type of locational conflict derives from public coordination *within* jurisdictions as larger institutions impose externalities on those subordinate to them. The impure nature of certain public goods provides a raft of opportunities for groups of residents to seek to maximize positive externalities such as 'good schools' and minimize the impact of negative externalities.

In the final section of Chapter One, Cox (1973: 14) establishes a 'relatively simple model of locational conflict in an urban context.' As indicated above, the core of the argument is that decision-making units – whether households, firms or other organizations – allocate resources to maximize their utility. The various externalities that result generate locational conflict between proximate decision-makers. Locational conflicts can be resolved through relocation or through private bargaining, although these solutions tend to be 'suboptimal.' Accordingly, some measure of collective control over property rights is required to ensure more optimal solutions. One of the most interesting and persuasive aspects of the core argument is its independence of any one particular scale of analysis. As Cox (1973: 15) argues, 'the collectivization of property rights and the vesting of them in a superordinate authority … creates new decision-making units on a new and larger geographical scale. These units in turn create externalities for other units at the same geographic scale or at

smaller geographic scales and a new cycle of locational conflict and conflict resolution is generated.'

The remainder of the book seeks to apply the conceptual model to conflicts within metropolitan areas. Chapter Two establishes the broad context and examines both the way in which populations are localized into neighborhoods and jurisdictions as well as how these entities are organized territorially within the city. In short, Cox argues that there are *de facto* and *de jure* forms of territorial organization. The *de facto* organization of the city is one based on neighborhoods which tend towards homogeneity in terms of income, ethnicity and race. The *de jure* organization then tends to regularize or institutionalize such segregation through jurisdictional fragmentation on the one hand and the central city/suburban divide on the other. Yet segregation into discrete neighborhoods cannot guarantee control of externalities, meaning that securing the provision of positive externalities commonly depends upon what Cox terms the 'respecification of property rights' in the form of the coercive power of governmental authority. The *de facto* and *de jure* forms of organization are routinely coupled and in Chapter Two Cox argues that the power to secure positive externalities while avoiding the negative is related to the extent to whether neighborhoods are 'organized' or not. Yet this is not the sole determinant of which locations get what with Cox suggesting, as a segue into Chapter Three, that interjurisdictional differences and relationships are also vitally important.

Chapter Three focuses specifically on what Cox terms the central city–suburban fiscal disparities problem. In some ways this is a peculiarly American phenomenon given the way in which these spatial disparities are institutionalized by the fragmentation of the city into jurisdictions that have considerable latitude in funding and decision-making. Over the past three decades a considerable amount of effort has been spent examining the nature and extent of divisions between US central cities and their suburbs and possible ways of addressing such disparities. Potential solutions include mechanisms such as fiscal transfer, metropolitan consolidation and other forms of regional governance. Cox's original account exposes the powerful economic logic that has generated and then reproduced city–suburb inequalities. The context here was the relative impoverishment of the central city, the aging of its housing stock, the fiscal problems created through the suburbanization of business and the selective suburbanization of wealthier white residents, otherwise known as 'white flight.'

One of the highlights of Chapter Three is Cox's quantitative evaluation of disparities, examining the extent and nature of their variability over space. Cox uses factor analysis to identify the metropolitan areas that exhibit the most egregious imbalances between 'need' or demand for public services and available tax resources. The analysis indicates 'the fiscal disparities problem is largely a problem of the US Northeast and Midwest' (Cox, 1973: 46). Cox attributes this to national migration patterns imposing costs or negative externalities on Northern central cities but also, and perhaps more significantly, to the locational arrangements of individuals and households within metropolitan areas. In brief, Cox argues wealthier residents are able to capitalize on the positive externalities of metropolitan living while escaping many of the negative externalities through suburban residence. The exclusive nature of the suburbs is maintained by a number of mechanisms. These include exclusionary zoning practices in the form of minimum lot sizes and zoning for single-family residences. We can see these practices as a form of NIMBYism (the 'not in my back yard' syndrome) in which those with wealth and resources that depend upon certain 'less desirable' land-uses, such as commercial and industrial developments, are nevertheless able to exclude them from their particular, immediate neighborhood.

Discrimination within housing and employment markets also plays a significant sorting role while the general disparity in property values between city and suburb further serves to exclude low-income residents. As Cox (1973: 59) suggests, 'those who can meet the requirements of the suburban filter suburbanize. Those who cannot meet the requirements stay in the city.' While there are problems here in terms of the apparent universal nature of such preferences – think of the way in which gentrification marked a return to the central city by those with economic assets (Ley, 1996) – the selective nature of suburbanization remains a powerful force underpinning contemporary socio-spatial disparities within US metropolitan areas.

Chapter Four shifts the scale of analysis to examine locational conflicts within local jurisdictions. This is an important complement to Chapter Three for, as Cox suggests, the elimination of metropolitan fragmentation would not necessarily eliminate inequities in public allocation and thus the basis for locational conflict. Variation in educational provision – especially within central cities – is once again seen as a critical source of externalities that reinforce existing urban inequalities. One of the most interesting aspects of Chapter Four is the way in which Cox makes clear the relationship between political power and inequalities in resource provision and the allocation of public goods. His discussion challenges the pluralist view that each vote carries equal weight in determining policy outcomes. Towards the end of Chapter Four, Cox examines in more detail two groups that exercise political leverage that greatly outweighs their electoral significance. The discussion of middle-class households reaffirms the general notion that those with economic assets are better able to secure preferable political outcomes. However, the discussion of 'the downtown business elite' hints that certain interests have an even more vital stake in securing positive flows of resources and value as a consequence of the immobility of their investments. Those familiar with debates on the local dependence of certain business and property-owners can clearly trace a line of argument from *Conflict, Power and Politics in the City* to the much cited 1988 *Annals of the Association of American Geographers* paper, where Cox develops an argument about the geographic fixity of particular local interests in a global era (Cox and Mair, 1988). The suggestion that urban inequalities are the product of the relationship between economic assets and political power has indeed been a vital one for subsequent developments in urban geography and urban studies more broadly.

Chapter Five examines the policy implications of Cox's model of metropolitan inequalities and the conflicts they generate. His is an explicitly redistributive argument and in this sense a liberal claim for social change in contrast to the more radical strategies favored by later Marxist accounts. Cox argues social injustice is the product of two basic conditions. The first is the spatial organization of the urban political system such that the benefits derived from the public allocation of resources 'depends very much on location relative to … job opportunities, schools, freeways, and the boundaries of different municipalities and of school catchment areas' (Cox, 1973: 106). But injustice is also seen as the product of a second basic inequality in the allocation of private resources, with those with economic power able to lobby and influence more effectively than those without. In short, Cox argues spatial inequity in public provision is 'strongly correlated with social inequity' (Cox, 1973: 126) and while his discussion of policy options is divided into those that effect changes in *spatial* organization and those that focus on the *social* allocation of income and wealth it is clear that the favored solution is some combination of a change in metropolitan form – such as metropolitan integration – combined with an effective redistribution of income through taxation. In his estimation, greater social and spatial equalization of resources should serve to mitigate or even eliminate negative externalities, undermine segregation and thus produce a more just and equitable city.

Evaluating *Conflict, Power and Politics in the City*

The strengths and limits of *Conflict, Power and Politics in the City* can be examined at two different levels. The first is in terms of the book's immediate reception; the second its more lasting impact. The initial reception of the book was decidedly positive. Murray Austin, in the *Annals of the Association of American Geographers*, referred to it as 'a major contribution to the study of urban problems' (Austin, 1973: 389); Melvin Albaum (1973) saw it as 'an important volume' in the McGraw-Hill series, while Norman Walzer described it as 'an illuminating analysis' (Walzer, 1973: 476). These initial reviews pointed to a number of clear strengths. The first is the unabashedly conceptual nature of Cox's contribution (Ironside, 1976). *Conflict, Power and Politics in the City* provided possibly the first specifically geographical framework for examining the city and its politics. In doing so it builds upon a set of key concepts and ideas in which geographical notions such as proximity, location, accessibility and scale are central. Over the years there have been many contributions to the geography literature that have sought to 'add' geography to an existing mix of concepts and claims. Such work has tended to spatialize ideas and theories from other disciplines rather than develop the geographer's conceptual toolbox. In contrast, the foundation of Cox's argument is a set of *geographical* concepts that are developed into an overarching framework that insists on the centrality of space in producing and resolving conflict in the city.

Second, it is apparent that the book blends conceptual rigour with a rich and varied array of empirical illustrations. These are drawn overwhelmingly from the US with one or two complementary examples from the UK (see below for the limits to such a strategy). But this is by no means a book that argues *by* empirical example. Instead Cox draws on a wide range of materials, including academic case studies, government reports such as Senate and Congressional records, and newspaper sources to present a range of cases that illustrate the broader conceptual framework and the geographical basis of urban politics. The examples are generally routine rather than exceptional but that serves as an even more effective illustration of the general purchase of Cox's framework. A third notable contribution of the text is that it clearly establishes the policy implications of the analysis.

The book also had a more lasting influence on a number of subsequent literatures. The first was on locational politics and the study of conflicts arising from externalities (O'Loughlin and Munski, 1979; Ley and Mercer, 1980). These themes have been of particular interest to urban geographers and the study of negative externalities and NIMBYism has proved to be especially enduring (Lake, 1987). A second set of geographical studies examined questions of inequality in access to public facilities and in the distribution of resources across the city (for a review see DeVerteuil, 2000). Thirdly, Cox's book was influential – largely beyond geography this time – in the study of the political activity of neighborhood groups (Rich, 1980).

While the book has a number of very positive attributes, with the benefit of hindsight, it is possible to identify four key limitations to Cox's text. Hindsight always provides perspective and many of the limitations have been brought into sharper relief by subsequent developments in human geography. The first – and arguably the most significant – is the book's reliance on an economistic public choice model centered on the rational, utility-maximizing individual. For Cox, individual residents necessarily maximize the return on their investments, calculate the costs and returns to alternative options and mechanistically follow the course given by the most rational return. In 1973 this was the dominant view in economics and Cox simply applied what he saw as 'intellectually intriguing' work to the question of locational allocation and conflict in the city. The limitations of such an approach are now well known – including its assumptions about the rational and self-interested nature of individual action – and even the initial reviews had started to question the viability of such an approach. As time has gone on the approach looks increasingly archaic and at a time when cities are once more rising up the academic agenda the social theories that underpin the normative claims of geographers such as Ash Amin (2006), David Harvey (2000) and Nigel Thrift (2005) could not be more different from the neoclassical model that formed the foundation for Cox's account.

If the first limitation derives from the book's commitment to what became a quite problematic theoretical framework then the second reflects the book's rather limited range. Again we should not see *Conflict, Power and Politics in the City* in isolation given that many contemporaneous texts also focused on the question of 'who gets what, where' (Castells, 1977 is the most prominent company here). From this perspective urban conflict is seen as conflict over the distribution of resources as well as the relative balance between costs and benefits in producing use values. In retrospect, Kevin Cox would see this as a major limitation and his turn to Marxism clearly affirmed the limitations of an approach that ignored questions of production and the generation of surplus. For those who were influenced by Marx, the social welfare question of 'who gets what' was displaced in the 1970s and early 1980s by a concern with how the 'what' is produced in the first place and by an analysis of the social relations forged in the sphere of production. For Cox himself, this shift involved less a rejection of the significance of the 'urban' and its politics and more an attempt to link urban conflicts and outcomes to class relationships rather than to those manifest in the neighborhoods, school districts and residents associations that are central to *Conflict, Power and Politics in the City*.

A third limitation similarly reflects the book's limited terms of reference. The essence of *Conflict, Power and Politics in the City* is an attractive but ultimately abstract account of the mechanisms producing conflict in the city. But of course *the* city is a particular one – the US city. The nature of the McGraw-Hill series in combination with Cox's own interests largely dictated the geographical limits of the book, but the failing here is less the focus on a single country – and, as noted above, there are examples drawn from Britain that offset that particular critique – and more that the book failed to recognize the particularities of the US case. The dominance of private property and the intense fragmentation of governmental authority are widespread conditions for sure, but they are not universal – Havana is not Guangzhou which is not Detroit – and in this rather obvious sense the book perhaps failed to talk to as wide an audience as possible.

Fourth, while the book is resolutely geographical the geographies are very clearly defined by notions of fixity, territory and bounded space. In some ways this is an artifact of Cox's focus on the neighborhood as the primary locus of social life and the claim that neighborhoods tend to be homogenous with respect to income and race. With this assumption Cox avoids questions relating to the dynamics of the household and especially gender relations (Drake and Horton, 1983), as well as the influence of the national and the transnational on the city. The 1970s marked a reversal in the long-term demographic trend of a decreasing proportion of foreign-born in the US population and in cities like Miami,

Los Angeles and New York the role of the transnational came to play a much more prominent role in the urban imaginary. This was partly the consequence of migration but also, perhaps, about the increasing integration of these cities into regional and global flows of investment, trade and ideas. While 'globalization' had yet to emerge as a significant academic concern, Cox's emphasis on the territorial, the bounded and the fixed has seemed to lose its resonance in a globalizing world in which the focus has shifted to the mobile, the mutable and the unbound.

Conclusion

Although Cox's book was never as widely cited as many of the other classic texts in this collection, the influence of *Conflict, Power and Politics in the City* endured well beyond the initial flurry of positive reviews. Reading it today it remains a simple, elegant and yet rigorous argument about the nature of conflict in the US city. Yet this is not a book that serves as a model statement of any one particular view, approach or era. Indeed its enduring legacy might best be seen as a transitional work between a spatial science tradition characterized by parsimony and simplicity and a much more politicized view of the world in which political influence and power are closely tied to economic assets and interests.

DeVerteuil (2000: 59) argues that Cox's work was 'instrumental in recasting facility location away from neoclassical, quantitative models normatively concerned with spatial outcomes and toward a more conflictual, political and socially embedded analytical framework concerned with spatial processes.' While DeVerteuil's review is focused on one aspect of *Conflict, Power and Politics in the City* his assessment of Cox's contribution could be fruitfully applied to a range of additional issues and themes. Indeed, the most enduring legacy of all was precisely this focus on the political nature of locational conflict and Cox's emphasis on the significance of government – or what we would later term 'the state' – in shaping the city. Subsequent work, not least his own, would develop these ideas in terms of the political-economic dynamics of the city but Cox's book insists that the city is the product of forces that derive their power from private property on the one hand and governmental or state authority on the other. These two sources of power, and the relationship between them, would form the basis for a great deal of subsequent work on the nature and dynamics of the city. Cox has subsequently made a substantial direct contribution to this field through

his work on the 'new urban politics' (see Cox, 1993), its relationship to globalization (Cox, 1995) as well as his work on the geographic specificity of politics and political conflict (Cox and Mair, 1988; Cox, 1998). Much of this work has continued to emphasize the relationship between state power and the realization of economic interests in the city. While the rejection of what had come before was by no means as complete or dramatic as Harvey's book of the same year (1973) the central focus on the relationship between private property and the exercise of political power surely marked a welcome turn away from avowedly apolitical location theory towards a thoroughly politicized urban geography.

Secondary sources and references

Albaum, M. (1973) 'Conflict, Power and Politics in the City' (book review), *Professional Geographer* 26 (3): 339.

Amin, A. (2006) 'The good city', *Urban Studies* 43 (5–6): 1009–1023.

Austin, C.M. (1973) 'Conflict, Power and Politics in the City' (book review), *Annals of the Association of American Geographers* 63 (3): 389–390.

Bach, W. (1972) *Atmospheric Pollution.* New York: McGraw-Hill.

Barnes, T.J. (2004) 'Placing ideas: genus loci, heterotopia and geography's quantitative revolution', *Progress in Human Geography* 28 (5): 565–595.

Castells, M. (1977) *The Urban Question: A Marxist Approach.* London: Edward Arnold.

Cox, K.R. (1973) *Conflict, Power, and Politics in the City: A Geographic View.* New York: McGraw-Hill.

Cox, K.R. (1993) 'The local and the global in the new urban politics – a critical view', *Environment and Planning D: Society and Space* 11 (4): 433–448.

Cox, K.R. (1995) 'Globalization, competition and the politics of local economic development', *Urban Studies* 32 (2): 213–224.

Cox, K.R. (1998) 'Spaces of dependence, spaces of engagement and the politics of scale, or: looking for local politics', *Political Geography* 17 (1): 1–23.

Cox, K.R. and Mair, A. (1988) 'Locality and community in the politics of local economic development', *Annals of the Association of American Geographers* 78 (2): 307–325.

DeVerteuil, G. (2000) 'Reconsidering the legacy of urban public facility location theory in human geography', *Progress in Human Geography* 24 (1): 47–69.

Drake, C. and Horton, J. (1983) 'Comment on editorial essay: sexist bias in political geography', *Political Geography Quarterly* 2 (4): 329–337.

Harvey, D. (1973) *Social Justice and the City.* London: Edward Arnold.

Harvey, D. (2000) *Spaces of Hope.* Berkeley: University of California Press.

Hirschman, A. (1970) *Exit, Voice and Loyalty.* Cambridge: Harvard University Press.

Ironside, R.G. (1976) 'Conflict, Power and Politics in the City' (book review), *Geographical Analysis* 8 (1): 104–109.

Lake, R. (1987) *Resolving Locational Conflict*. New Brunswick, NJ: Center for Urban Policy Research.

Ley, D. (1974) *The Inner City Ghetto as Frontier Outpost*. Washington, DC: Association of American Geographers.

Ley, D. (1996) *The New Middle Class and the Remaking of the Central* City. Oxford: Oxford University Press.

Ley, D. and Mercer, J. (1980) 'Locational conflict and the politics of consumption', *Economic Geography* 56 (2): 89–109.

Morrill, R. and Wohlenberg, E. (1971) *The Geography of Poverty in the United States*. New York: McGraw-Hill.

O'Loughlin, J. and Munski, D.C. (1979) 'Housing rehabilitation in the inner city – comparison of 2 neighborhoods in New Orleans', *Economic Geography* 55 (1): 52–70.

Rich, R.C. (1980) 'A political-economy approach to the study of neighborhood organizations', *American Journal of Political Science* 24 (4): 559–592.

Smith, D. (1973) *The Geography of Social Well-being in the United States*. New York: McGraw-Hill.

Thrift, N. (2005) 'But malice aforethought: cities and the natural history of hatred', *Transactions of the Institute of British Geographers* 30 (2): 133–150.

Walzer, N. (1973) 'Conflict, Power and Politics in the City' (book review), *Journal of Regional Science* 13 (3): 473–477.

Ward, D. (1971) *Cities and Immigrants*. New York: Oxford University Press.

6 PLACE AND PLACELESSNESS (1976): EDWARD RELPH

David Seamon and Jacob Sowers

A deep human need exists for associations with significant places. If we choose to ignore that need, and to allow the forces of placelessness to continue unchallenged, then the future can only hold an environment in which places simply do not matter. If, on the other hand, we choose to respond to that need and to transcend placelessness, then the potential exists for the development of an environment in which places are for [people], reflecting and enhancing the variety of human experience. Which of these two possibilities is most probable, or whether there are other possibilities, is far from certain. But one thing at least is clear – whether the world we live in has a placeless geography or a geography of significant places, the responsibility for it is ours alone. (Relph, 1976: 147)

Introduction

Geographers have long spoken of the importance of *place* as the unique focus distinguishing geography from other disciplines. Astronomy has the heavens, History has time, and Geography has place. A major question that geographers must sooner or later ask, however, is 'What exactly is place?' Is it merely a synonym for location, or a unique ensemble of nature and culture, or could it be something more?

Beginning in the early 1970s, geographers such as Yi-Fu Tuan (1974), Anne Buttimer (1976), and Edward Relph (1976, 1981, 1993) grew dissatisfied with what they felt was a philosophically and experientially anemic definition of place. These thinkers, sometimes called 'humanistic geographers,' probed place as it plays an integral role in human experience. One influential result of this new approach was Edward Relph's *Place and Placelessness,* a book that continues to have significant conceptual and practical impact today, both inside and outside geography.

In the early 1970s, Relph was a doctoral student at the University of Toronto, working on his dissertation concerning the relationship between Canadian national identity and the symbolic landscapes of the Canadian Shield, especially those represented by lakes and forests (Relph, 1996). As his project progressed, he became dissatisfied with the lack of philosophical sophistication given to the definition of place. Relph found this supposed conceptual pillar of the discipline to be superficial and incomplete, especially in terms of the importance of place in ordinary human life. How could one study place attachment, sense of place, or place identity without a clear understanding of the depth and complexity of place as it is experienced and fashioned by real people in real places? Eventually, Relph scrapped his Canadian Shield study and shifted focus to a broader look at the nature and meaning of place as it plays an integral part in the lives of human beings.

A phenomenology of place and space

Published in 1976, *Place and Placelessness* is a substantive revision of Relph's 1973 University of Toronto doctoral dissertation in Geography. As he emphasizes at the start of the book, his research method is 'a phenomenology of place' (Relph, 1976: 4–7). Phenomenology is the interpretive study of human experience. The aim is to examine and to clarify human situations, events, meanings, and experiences as they are known in everyday life but typically unnoticed beneath the level of conscious awareness (Seamon, 2000). One of phenomenology's great strengths is seeking out what is obvious but unquestioned and thereby questioning it. To uncover the obvious, we must step back from any taken-for-granted attitudes and assumptions, whether in the realm of everyday experience or in the realm of conceptual perspectives and explanations, including the scientific. In *Place and Placelessness*, Relph steps back to call into question the taken-for-granted nature of place and its significance as an inescapable dimension of human life and experience.

Relph begins *Place and Placelessness* with a review of space and its relationship to place. He argues that space is not a void or an isometric plane or a kind of container that holds places. Instead, he contends that, to study the relationship of space to a more experientially-based understanding of place, then space too must be explored in terms of how people experience it. Although Relph says that there are countless types and intensities of spatial experience, he delineates a heuristic structure grounded in 'a continuum that has direct experience at one extreme and abstract thought at the other...' (Relph, 1976: 9). On one hand, he identifies modes of spatial experience that are instinctive, bodily, and immediate – for example, what he calls pragmatic space, perceptual space, and existential space. On the other hand, he identifies

modes of spatial experience that are more cerebral, ideal, and intangible – for example, planning space, cognitive space, and abstract space. Relph describes how each of these modes of space-as-experienced has varying intensities in everyday life. For example, existential space – the particular taken-for-granted environmental and spatial constitution of one's everyday world grounded in culture and social structure – can be experienced in a highly self-conscious way such as when one is overwhelmed by the beauty and sacredness of a Gothic cathedral; or in a tacit, unself-conscious way as one sits in the office day after day paying little attention to his or her surroundings.

Although the spatial modes that Relph identifies may each play a particular role in everyday experience, Relph emphasizes that in reality these modes are not mutually exclusive but all part and parcel of human spatial experience as a lived, indivisible whole. For example, he explains that cognitive conceptions of space understood through maps may help to form our perceptual knowledge, which in turn may color our day-to-day spatial encounters as we move through real-world places. Though a radical idea in the 1970s, Relph's conclusion that space is heterogeneous and infused with many different lived dimensions is largely taken for granted in geographical studies today as researchers speak of such spatial modes as sacred space, gendered space, commodified space, and the like.

One of Relph's central accomplishments in *Place and Placelessness* is his preserving an intimate conceptual engagement between space and place. Many geographers speak of both concepts but ultimately treat the two as separate or give few indications as to how they are related existentially and conceptually. For Relph, the unique quality of place is its power to order and to focus human intentions, experiences, and actions spatially. Relph thus sees space and place as dialectically structured in human environmental experience, since our

understanding of space is related to the places we inhabit, which in turn derive meaning from their spatial context.

Depth of place

A central reason for Relph's exhaustive study of place is his firmly held belief that such understanding might contribute to the maintenance and restoration of existing places and the making of new places (also see Relph, 1981, 1993). He argues that, without a thorough understanding of place as it has human significance, one would find it difficult to describe why a particular place is special and impossible to know how to repair existing places in need of mending. In short, before we can properly prescribe, we must first learn how to accurately describe – a central aim of phenomenological research.

In examining place in depth, Relph focuses on people's identity *of* and *with* place. By the identity *of* a place, he refers to its 'persistent sameness and unity which allows that [place] to be differentiated from others' (Relph, 1976: 45). Relph describes this persistent identity in terms of three components: (1) the place's physical setting; (2) its activities, situations, and events; and (3) the individual and group meanings created through people's experiences and intentions in regard to that place.

Relph emphasizes, however, that place identity defined in this threefold way is not sufficiently pivotal or deep existentially because, most essentially, places are 'significant centres of our immediate experiences of the world' (Relph, 1976: 141). If places are to be more thoroughly understood, one needs a language whereby we can identify particular place experiences in terms of the intensity of meaning and intention that a person and place hold for each other. For Relph, the crux of this lived intensity is identity *with* place, which he defines through the concept of *insideness* – the degree of attachment, involvement, and concern that a person or group has for a particular place.

Insideness and outsideness

Relph's elucidation of insideness is perhaps his most original contribution to the understanding of place because he effectively demonstrates that this concept is the core lived structure of place as it has meaning in human life. If a person feels inside a place, he or she is here rather than there, safe rather than threatened, enclosed rather than exposed, at ease rather than stressed. Relph suggests that the more profoundly inside a place a person feels, the stronger will be his or her identity with that place.

On the other hand, a person can be separate or alienated from place, and this mode of place experience is what Relph calls *outsideness*. Here, people feel some sort of lived division or separation between themselves and world – for example, the feeling of homesickness in a new place. The crucial phenomenological point is that outsideness and insideness constitute a fundamental dialectic in human life and that, through varying combinations and intensities of outsideness and insideness, different places take on different identities for different individuals and groups, with human experience taking on different qualities of feeling, meaning, ambience, and action.

The strongest sense of place experience is what Relph calls *existential insideness* – a situation of deep, unself-conscious immersion in place and the experience most people know when they are at home in their own community and region. The opposite of existential insideness is what he labels *existential outsideness* – a sense of strangeness and alienation, such as that often felt by newcomers to a place or by people who, having been away from their birth place, return

to feel strangers because the place is no longer what it was when they knew it earlier.

In his book, Relph discusses seven modes of insideness and outsideness (no doubt there are more) grounded in various levels of experiential involvement and meaning. The value of these modes, particularly for self-awareness, is that they apply to specific place experiences yet provide a conceptual structure in which to understand those experiences in broader, more explicit terms.

Placelessness

In the last half of the book, Relph examines ways in which places may be experienced *authentically* or *inauthentically* (terms borrowed from phenomenological and existential philosophy). An *authentic* sense of place is 'a direct and genuine experience of the entire complex of the identity of places – not mediated and distorted through a series of quite arbitrary social and intellectual fashions about how that experience should be, nor following stereotyped conventions' (Relph, 1976: 64).

Individuals and groups may create a sense of place either unself-consciously or deliberately. Thus, because of constant use, a nondescript urban neighborhood can be as authentic a place as Hellenic Athens or the Gothic cathedrals – the latter both examples, for Relph, of places generated consciously. Relph argues that, in our modern era, an authentic sense of place is being gradually overshadowed by a less authentic attitude that he called *placelessness*: 'the casual eradication of distinctive places and the making of standardized landscapes that results from an insensitivity to the significance of place' (Relph, 1976: Preface).

Relph suggests that, in general, placelessness arises from *kitsch* – an uncritical acceptance of mass values, or *technique* – the overriding concern with efficiency as an end in itself. The overall impact of these two forces, which manifest through such processes as mass communication, mass culture, and central authority, is the 'undermining of place for both individuals and cultures, and the casual replacement of the diverse and significant places of the world with anonymous spaces and exchangeable environments' (Relph, 1976: 143).

Influence of *Place and Placelessness*

Since Relph's book was published, there has been a spate of popular studies on the nature of place. In addition, thinkers from a broad range of conceptual perspectives – from positivist and neo-Marxist to post-structuralist and social-constructivist – have drawn on the idea of place, though understanding it in different ways and using it for different theoretical and practical ends (Creswell, 2004; Seamon, 2000).

Scholarly interest in *Place and Placelessness* has steadily increased over the years. According to citation indices in the sciences, social sciences, and the arts and humanities, the book has been referenced in scholarly journals a total of 357 times from 1977 to 2005. In the first 10 years, there was an average of some 12 citations per year; since then, references have steadily increased to 36 entries in 2004. Geographers have cited the book most since 1989 (142 entries), though scholars in environmental studies also demonstrate strong interest (118 entries). In addition, the book has been cited by researchers in psychology (43 times), sociology (42), urban studies (30), planning (21), health (10), and anthropology (9).

To provide the reader with an indication of how Relph's ideas in *Place and Placelessness* have been used as a major conceptual mooring point by other researchers, we highlight three examples – one book, one article, and one dissertation (for a more extensive list, see Seamon, 2000). Published two years after Relph's book, geographer David Seamon's *A Geography of the Lifeworld* is the first major study to draw on Relph's notion of insideness

and to demonstrate how it could be extended phenomenologically to examine a topic that Seamon calls *everyday environmental experience* – the sum total of people's firsthand involvements with the geographical world in which they live (Seamon, 1979: 15–16). Seamon considers how, through experienced dimensions like body, feelings, and thinking, the quality of insideness is expressed geographically and environmentally. Seamon's work illustrates how Relph's phenomenology of place offers a field of conceptual clarity from which other researchers might embark on their own phenomenological explorations.

A second study illustrating the conceptual potential of *Place and Placelessness* is landscape architect V. Frank Chaffin's research, which focuses on Isle Brevelle, a 200-year-old river community on the Cane River of Louisiana's Natchitoches Parish (Chaffin, 1989). Through an interpretive reading of the region's history and geology, in-depth interviewing of residents, and his own personal encounters with Isle Brevelle's landscape while canoeing on the Cane River, Chaffin aims to reach an *empathetic insideness* with this place – in other words, he attempts to find ways to be open to and thereby to understand more deeply Isle Brevelle's unique sense of place. One central aspect of Chaffin's encounter with this place is his unexpected realization that the Cane River is not an edge that separates its two banks but, rather, a seam that gathers the two sides together as one community and one place.

A third study using themes from *Place and Placelessnesss* for conceptual mooring is psychologist Louise Million's dissertation (Million, 1992), which examines phenomenologically the experience of five rural Canadian families forced to leave their ranches because of the construction of a reservoir dam in southern Alberta. Drawing on Relph's modes of insideness and outsideness, Million identifies the central lived qualities of what she calls *involuntary displacement* – the families' experience of forced relocation and resettlement. Making use of in-depth interviews with the five families, she demonstrates how place is prior to involuntary displacement, with the result that this experience can be understood existentially as a forced journey marked by eight stages – (1) *becoming uneasy*, (2) *struggling to stay*, (3) *having to accept*, (4) *securing a settlement*, (5) *searching for the new*, (6) *starting over*, (7) *unsettling reminders*, and (8) *wanting to resettle*. In delineating the lived stages in the process of losing place and attempting to resettle, Million's study demonstrates how Relph's modes of insideness and outsideness can be used developmentally to examine place experience and identity as they strengthen, weaken, or remain more or less continuous over time.

Criticisms of *Place and Placelessness*

Broadly, one finds three major criticisms of *Place and Placelessness*: that it is essentialist; out of touch with what places really are today; and structured around simplistic dualisms that misrepresent and limit the range of place experience, particularly the possibility of a 'global sense of place' (Massey, 1997: 323). The essentialist claim has been brought forth especially by Marxists (e.g., Peet, 1998: 63) and social constructivists (e.g., Cresswell, 2004: 26, 30–33), who argue that Relph presupposes and claims an invariant and universal human condition that will be revealed only when all 'non-essentials,' including historical, cultural, and personal qualities, are stripped away, leaving behind some inescapable core of human experience. These critics point out that, in focusing on the experience of place as a foundational existential quality and structure, Relph ignores specific temporal, social, and individual circumstances that shape particular places and particular individuals' and groups' experience of them.

This criticism misunderstands the basic phenomenological recognition that there are different dimensions of human experience and existence that *all must be incorporated* in a thorough understanding of human and societal phenomena. These dimensions include: (a) one's unique personal situation – e.g., one's gender, physical and intellectual endowments, degree of ableness, and personal likes and dislikes; (b) one's unique historical, social, and cultural situation – e.g., the era and geographical locale in which one lives, his or her economic and political circumstances, and his or her educational, religious, and societal background; and (c) one's situation as a typical human being who sustains and reflects a typical human world – e.g., Relph's claim that place is an integral lived structure in human experience.

What is exciting and dynamic about Relph's broad conclusions regarding place – which first of all relate to dimension (c) – is their potential as starting points for more specific phenomenological investigations of (a) and (b) as exemplified, for example, in the real-world studies of Chaffin and Million highlighted above. Chaffin's research demonstrates how Relph's broad principles can inform and direct phenomenological research focusing on the social and cultural dimensions of one specific place – the Cain River community. Similarly, in her study of the displaced Alberta ranchers, Million illustrates how Relph's general principles and conclusions can guide empirical research in regard to specific individuals and families in a specific place, time, and situation. In turn, Chaffin and Million's more grounded discoveries clarify and amplify Relph's broader claims.

In short, Relph's phenomenology of place points toward a conceptual and methodological reciprocity between the general and the specific, between the foundational and the particular, between the conceptual and the lived. This convincing 'fit' among levels is a hallmark of the best phenomenology.

A lack of conceptual sophistication?

In a commentary written for *Place and Placelessness'* 20th anniversary, Relph (1996) suggested that, in hindsight, another major weaknesses of *Place and Placelessness* was its lack of conceptual sophistication, particularly its straightforward use of dialectical opposites as a way to conceptualize place experience – insideness/outsideness, place/placelessness, authenticity/inauthenticity, and so forth. One result is that critics have often misunderstood Relph's point of view, claiming he favored places over placelessness, insideness over outsideness, authentic over inauthentic places, rootedness over mobility, and place as a static, bounded site over place as a dynamic, globally-connected process (Cresswell, 2004; Massey, 1997; Peet, 1998).

If, however, one reads the book carefully and draws on his or her own personal experiences of place for evidence and clarification, he or she realizes the extraordinary coverage and flexibility of Relph's conceptual structure. Especially through the continuum of insideness and outsideness, he provides a language that allows for a precise designation of the particular experience of a particular person or group in relation to the particular place in which they find themselves. Relph also provides a terminology for describing how and why the same place can be experienced differently by different individuals (e.g., the long-time resident vs. the newcomer vs. the researcher who studies the place) or how, over time, the same person can experience the same place differently at different times (e.g., the home and community that suddenly seem so different when one's significant other dies).

As Relph's book strikingly demonstrates, a major strength of phenomenological insights is their provision of a conceptual language that allows one to separate from taken-for-granted everyday experience – the *lifeworld* as it is called phenomenologically (Buttimer, 1976; Seamon, 1979, 2000). Too often, researchers lose sight of the need to move outside lifeworld descriptions and terminology, and the result is confusion or murkiness as to the exact phenomenon they are attempting to understand.

For example, in feminist and cultural-studies research that focuses on negative and traumatic images of place (e.g., Rose, 1993: 53–55), an emphasis is sometimes given to how family violence generates homes where family members feel victimized and insecure. Too often, the post-structural and social-constructivist conclusion is to call into question the entire concept of home and place and to suggest that they might be nostalgic, essentialist notions that need vigorous societal and political modification – perhaps even substitution – in postmodern society.

Relph's modes of insideness and outsideness point to an alternative understanding. The problem is not home and place but a conceptual conflation for which Relph's language provides a simple corrective: the victim's experience should not be interpreted as a lack of at-homeness but, rather, as one mode of existential outsideness, which in regard to one's most intimate place – the home – is particularly undermining and potentially life-shattering.

Relph's notion of existential outsideness allows us to keep the experiences of home and violation distinct. Through his lived language of place, we can say more exactly that domestic violence, whether in regard to women *or* men, is a situation where a place that typically fosters the strongest kind of existential insideness has become, paradoxically, a place of overwhelming existential outsideness. The lived result must be profoundly destructive.

The short-term phenomenological question is how these victims can be helped to regain existential insideness. The longer-term question is what qualities and forces in our society lead to a situation where the existential insideness of home and at-homeness devolves into hurtfulness and despair. Something is deeply wrong, and one cause of the problem

may be the very problem itself – i.e., the growing disruption and disintegration of places and insideness at many different scales of experience, from home to neighborhood to city to nation (Fullilove, 2004; Relph, 1993).

How today to have insideness and place when change is constant, society is diverse, and so many of the traditional 'truths' no longer make sense is one of the crucial questions of our age. *Place and Placelessness* offers no clear answer, but it does provide an innovative language for thinking about the question.

Dwelling and journey

Another concern that some critics voiced regarding *Place and Placelessness* is that it favors home, center, and dwelling over horizon, periphery, and journey (Cresswell, 2004; Massey, 1997; Peet, 1998). As Relph (1996) says in his 20th-anniversary commentary, he was accused of emphasizing the positive qualities of place and ignoring or minimizing negative qualities – e.g., the possibility that place can generate parochialism, xenophobia, and narrow-mindedness (also see Relph, 2000). Again, a close reading of the book reveals a flexibility of expression – a recognition that an excess of place can lead to a provincialism and callousness for outsiders just as an excess of journey can lead to a loss of identity or an impartial relativity that allows for commitment to nothing. The broader point is that, in the book's lived dialectics (center/horizon, place/placelessness, and so forth), there is a wonderful resilience of conceptual interrelationship that is another hallmark of the best phenomenology.

In his 20th-anniversary commentary, Relph (1996) also points out that some critics mistakenly read the book as a nostalgic paean to pre-modern times and places (e.g., Peet, 1998). How could the kind of authentic places that he emphasized exist in our postmodern times of cyberspace, continuous technological change, human diversity, and geographical and social mobility?

This criticism, of course, ignores a central conclusion of *Place and Placelessness*: that regardless of the historical time or the geographical, technological, and social situation, *people will always need place* because having and identifying with place are integral to what and who we are as human beings (Casey, 1993; Malpas, 1999). From this point of view, the argument that postmodern society, through technological and cultural correctives, can now ignore place is questionable existentially and potentially devastating practically, whether in terms of policy, design, or popular understanding (Relph, 1993).

Instead, the crucial question that both theory and practice should ask is how a 'progressive' sense of place and insideness can be made *even in the context of* our relativist, constantly-changing postmodern world (Cresswell, 2004; Horan, 2000; Massey, 1997). Twenty years ago, Relph was one of the first thinkers to broach this question, which he explores in greater detail in his later *Rational Landscapes and Humanistic Geography* (Relph, 1981). Today, due to Relph's penetrating insights and the work of a small coterie of thinkers and practitioners like Christopher Alexander (2002–2005), Mindy Fullilove (2004), Bill Hillier (1996), Thomas Horan (2000), Daniel Kemmis (1995), and Robert Mugerauer (1994), we have the start of an answer to this question phrased in a phenomenological language (Seamon, 2004) that interprets place in a way considerably different from the post-structural, social-constructivist, and neo-Marxist perspectives that currently dominate academic discourse on place (Cresswell, 2004).

In spite of the dramatic societal and environmental changes that our world faces today, place continues to be significant both as a vigorous conceptual structure as well as an irrevocable part of everyday human life (Horan, 2000). This is not to suggest that the world must or could return to a set of distinct places all different, unconnected, and more or less unaware of each other. In today's globally-linked society, place independence is in many ways impossible (Cresswell, 2004; Relph, 2000). More so, the importance of place and locality must be balanced with an awareness of and connections to other places and global needs (Massey, 1997). The point is that an empathetic and compassionate understanding of the worlds beyond our own places may be best grounded in a love of a particular place to which I myself belong. In this way, we may recognize that what we need in our everyday world has parallels in the worlds of others (Relph, 1981, 1993).

Conclusion

Place and Placelessness is a remarkable demonstration of the potential conceptual and practical power of place, which, by its very nature, gathers worlds spatially and environmentally, marking out centers of human action, intention, and meaning that, in turn, help make place (Casey, 1993; Malpas, 1999). In many ways, the continuing dissolution of places and insideness in the world helps to explain the escalating erosion of civility and civilization, in the West and elsewhere. Relph's *Place and Placelessness* first pointed to this dilemma some 30 years ago and is today more relevant than ever.

Secondary sources and references

Alexander, C. (2002–2005) *The Nature of Order* (4 volumes). Berkeley, California: Center for Environmental Structure.

Buttimer, A. (1976) 'Grasping the dynamism of lifeworld', *Annals of the Association of American Geographers* 66: 277–292.

Casey, E. (1993) *Getting Back into Place: Toward a Renewed Understanding of the Place-World.* Bloomington: Indiana University Press.

Chaffin, V.F. (1989) 'Dwelling and rhythm: The Isle Brevelle as a landscape of home', *Landscape Journal* 7: 96–106.

Cresswell, T. (2004) *Place: A Short Introduction.* London: Blackwell.

Fullilove, M. (2004) *Root Shock.* New York: Ballantine.

Hillier, B. (1996) *Space is the Machine.* Cambridge: Cambridge University Press.

Horan, T.A. (2000) *Digital Places: Building Our City of Bits.* Washington, DC: Urban Land Institute.

Kemmis, D. (1995) *The Good City and the Good Life.* New York: Houghton Mifflin.

Malpas, J.E. (1999) *Place and Experience: A Philosophical Topography.* Cambridge: Cambridge University Press.

Massey, D. (1997) 'A global sense of place', in T. Barnes and D. Gregory (eds) *Reading Human Geography.* London: Arnold, pp. 315–323.

Million, M.L. (1992) *'It Was Home': A Phenomenology of Place and Involuntary Displacement as Illustrated by the Forced Dislocation of Five Southern Alberta Families in the Oldman River Dam Flood Area.* Doctoral dissertation, Saybrook Institute Graduate School and Research Center, San Francisco, California.

Mugerauer, R. (1994) *Interpretations on Behalf of Place: Environmental Displacements and Alternative Responses.* Albany, NY: State University of New York Press.

Peet, R. (1998) *Modern Geographic Thought.* Oxford: Blackwell.

Relph, E. (1976) *Place and Placelessness.* London: Pion.

Relph, E. (1981) *Rational Landscapes and Humanistic Geography.* New York: Barnes & Noble.

Relph, E. (1993) 'Modernity and the Reclamation of Place', in D. Seamon, (ed.) *Dwelling, Seeing, and Designing: Toward a Phenomenological Ecology.* Albany, NY: SUNY Press, pp. 25–40.

Relph, E. (1996) 'Reflections on *Place and Placelessness*', *Environmental and Architectural Phenomenology Newsletter* 7 (3): 14–16 [special issue on the twentieth anniversary of the publication of *Place and Placelessness*; includes commentaries by Margaret Boschetti, Louise Million, Douglas Patterson, and David Seamon].

Relph, E. (2000) 'Author's Response: *Place and Placelessness* in a New Context [Classics in Human Geography Revisited, *Place and Placelessness*]', *Progress in Human Geography* 24 (4): 613–619 [includes commentaries by John R. Gold and Mathis Stock].

Rose, G. (1993) *Feminism and Geography.* Cambridge: Polity.

Seamon, D. (1979) *A Geography of the Lifeworld: Movement, Rest, and Encounter.* New York: St. Martin's.

Seamon, D. (2000) 'A way of seeing people and place: Phenomenology in environment-behavior research', in S. Wapner, J. Demick, T. Yamamoto, and H. Minami (eds) *Theoretical Perspectives in Environment-Behavior Research.* New York: Plenum, pp. 157–178.

Seamon, D. (2004) 'Grasping the dynamism of urban place: Contributions from the work of Christopher Alexander, Bill Hillier, and Daniel Kemmis', in T. Mels (ed.) *Reanimating Places.* Burlington, VT: Ashgate, pp. 123–145.

Tuan, Y.-F. (1974) *Topophilia: A Study of Environmental Perceptions, Attitudes, and Values.* Englewood Cliffs, NJ: Prentice-Hall.

7 SPACE AND PLACE (1977): YI-FU TUAN

Tim Cresswell

Abstract knowledge about a place can be acquired in short order if one is diligent … But the 'feel' of a place takes longer to acquire. (Tuan, 1977: 183)

Introduction

I distinctly recall reading Yi-Fu Tuan's *Space and Place* as an undergraduate at University College London in 1985. I was taking a second year class called Humanistic Geography (changed a few years later to Cultural Geography – a sign of the times). It had been eight years since it was published and it had already become a classic text – a worthy follow-on to Tuan's 1974 book, *Topophilia*. I was immediately captured by it. It was like nothing I had ever read in a geography class. It was full of ideas but engagingly written. It was not obviously geography (as I understood it at the time) but yet everything about it seemed central to how geography could or even should be. It didn't have long sections on methodology or a review of recent literature. There was nothing pedantic about it. It just seemed to jump straight in and get on with thinking about some difficult questions.

I had no idea then how significant *Space and Place* would become to human geography and to me personally. By 1985 the idea of humanistic geography was a part of every undergraduate degree. It is fair to say that it had ceased to be revolutionary and had become simply a part of the way significant

parts of human geography were conducted. In 1977, when *Space and Place* was published, Humanistic Geography had been on most people's intellectual horizons for only around five years. The first book I ever bought for a geography course was Ley and Samuels' edited collection *Humanistic Geography: Prospects and Problems,* published in 1978. Edited collections such as this rely on a number of authors who can subscribe to the book's central idea and the willingness of a publisher to believe there is a market for it. That book, therefore, marked a degree of acceptance that something called Humanistic Geography had arrived. *Topophilia* had been published in 1974 and in the same year Tuan had published a paper called *Space and Place: Humanistic Perspective.* The other key texts in my 1985 course included work by Edward Relph (1976), Anne Buttimer and David Seamon (1980), David Ley (1974), Donald Meinig (1979) and J.B. Jackson (1980). Central to the work of all of these scholars was (and for many still is) the question of how people create a meaningful world and meaningful lives in the world. The notion of place was central to this endeavour.

The decade before *Space and Place* was published had seen human geography become something less than human. Spatial science, the quantitative revolution, and logical positivism had all sought to treat the world, and the people in it, as objects rather than subjects. The idea of rational people making rational choices in a rational world had held sway (see Abler *et al.,* 1971 and Haggett, 1965). Words like 'location', 'spatial patterns', 'distance' and 'space' had been central in the pages of books and journals. Human geography had operated more or less as a pseudo-science. Humanistic geography was, in part, a critique of this way of thinking about the world and the human inhabitation of it. The final pages of *Space and Place* make this critique clear. Although the book does not engage spatial science directly it does frequently compare the richness of an experiential perspective to more scientific and dehumanized approaches. 'What we cannot say in an acceptable scientific language we tend to deny or forget. A geographer speaks as though his knowledge of space and place were derived exclusively from books, maps, aerial photographs, and structured field surveys. He [sic] writes as though people were endowed with mind and vision but no other sense with which to apprehend the world and find meaning in it. He [sic] and the architect-planner tend to assume familiarity – the fact that we are oriented in space and home in place – rather then describe and try to understand what "being-in-the-world" is truly like' (1977: 200–201). And later, 'The simple being, a convenient postulate of science and deliberate paper figure of propaganda, is only too easy for the man [sic] in the street – that is, most of us – to accept' (1977: 203). This 'simple being' possibly refers to the abstract 'rational man' of spatial science and economics: the man who weighs up all options before making a rational choice about what to do next. Such a view of humanity has no place for meaning.

But *Space and Place* is not first and foremost a critique of spatial science. Its message is, for the most part, a positive one which asks us as geographers to be more aware of the ways in which we inhabit and experience the world – to increase, as Tuan puts it in the final line of the book, our 'burden of awareness'. Central to this awareness is the concept of place. The humanistic conception of place describes a way of relating to the world. Just as human geographers now are constantly

learning from theory and philosophy, often from continental Europe, so in the 1970s geographers looked to theories and philosophies for inspiration. The ideas that became most central to the humanistic endeavour were phenomenology and existentialism. These philosophies, put very simply, insisted that people had the burden of making their own meaning in the world through their own actions. In addition, the production of meaning arose from a process known as 'intentionality'. Intentionality described the relationship between human consciousness and the objects that people were conscious of. It is impossible to be conscious, phenomenology insisted, without being conscious of something. It is in this ofness that meaning was produced. Humanistic geographers, in general, but particularly Tuan, did not spend a great deal of time repeating the ideas and words of the philosophers who inspired them. Their task was not to explain the thoughts of others but to use these thoughts in the production of new knowledge. Tuan does, briefly, nod to the work of Paul Ricoeur in the early pages of *Space and Place* in order to underline the importance of the word 'experience' to his book.

Experience is directed to the external world. Seeing and thinking clearly reach out beyond the self. Feeling is more ambiguous. As Paul Ricoeur put it, 'Feeling is … without doubt intentional: it is a feeling of "something" – the lovable, the hateful [for instance]. But it is a very strange intentionality which on the one hand designates qualities felt on things, on persons, on the world, and on the other hand manifests and reveals the way in which the self is inwardly affected'. In feeling 'an intention and an affection coincide in the same experience' (Tuan quoting Ricoeur, 1977: 9).

Ideas such as 'experience' and 'feeling' were, by and large, not in the vocabulary of human geographers in the early 1970s. Spatial scientists were not very interested in how people related to the world through experience. Theirs was a world of simple people, meaning Tuan's ideas were revolutionary stuff. While the spatial scientists wanted to understand the world and treated people as part of that world (just like rocks, or cars or ice but with the magic ingredient of rationality added), Tuan, in *Space and Place*, focuses on the relationship between people and the world through the realm of experience. Tuan writes '[t]he given cannot be known in itself. What can be known is a reality that is a construct of experience, a creation of feeling and thought' (Tuan, 1977: 9). The focus of *Space and Place* is thus how we, as humans, are in-the-world – how we relate to our environment and make it into place.

The experience, as both thought and feeling, referred to in the book's subtitle, is central to the main argument of the book – the differentiation of space and place. What experience does is transform a relatively abstract notion of space into a relatively lived and meaningful notion of place. While space was the favoured object of the spatial scientist (and is still the favoured object of social theorists) it is the way space becomes endowed with human meaning and is transformed into place that lies at the heart of humanistic geography. 'What begins as undifferentiated space becomes place as we get to know it better and endow it with value … the ideas 'space' and 'place' require each other for definition. From the security and stability of place we are aware of the openness, freedom and threat of space, and vice versa. Furthermore, if we think of space as that which allows movement, then place is pause; each pause in movement makes it possible for location to be transformed into place' (Tuan, 1977: 6). This is the most important contribution of *Space and Place* to human geography – the distinction between an abstract realm of space and an experienced and felt world of place. Tuan forced geographers to stop taking these words for granted and to consider what they might mean for being human.

In addition to the space/place distinction which lies at the heart of the book there are many additional themes developed by Tuan that have since become central to human geography. Indeed, re-reading *Space and Place* recently, I was struck by how Tuan foreshadows and informs much of the most exciting work in the contemporary social sciences and humanities. While the notions of space and place Tuan developed have become commonsense (the ultimate compliment), notions of experience, the sensual, the emotional and the embodied are all the objects of much heated debate in geography departments today. Non-representational theory, for instance, as an examination of the pre-conceptual, mirrors Tuan's early musings on the centrality of experience and emotion (see Thrift, 2004; Lorimer, 2005). Back in 1977 Tuan was keen to reflect on the multiplicity of senses (not just sight) to understand our experience of the world in a fuller way. 'An object or place', Tuan writes, 'achieves concrete reality when our experience of it is total, that is, through all the senses as well as with the active and reflective mind' (1977: 18). Chapter Four is devoted to the ways in which space and place are understood in thoroughly embodied ways. Perhaps as a reaction to the thoroughly disembodied notion of rational economic man in spatial science, Tuan asserts that 'Man and world denote complex ideas. At this point, we also need to look at simpler ideas abstracted from man and world, namely, body and space, remembering however that the one not only occupies the other but commands and orders it through intention. Body is "lived body" and space is humanly constructed space' (1977: 35). The importance of the lived body is particularly underlined in the chapter on 'Spatial Ability, Knowledge and Place' (Chapter Six), Here Tuan prefigures recent work on the habitual and the pre-cognitive as he writes: 'There are many occasions on which we perform complex acts without the help of mental or material plans. Human fingers are exceptionally dexterous. A professional typist's fingers fly over the machine; all we see is blur of movement. Such speed and accuracy suggest that the typist knows the keyboard in the sense that he can envisage where all the letters are. But he cannot; he has difficulty recalling the positions of the letters that his fingers know so well' (1977: 68). Indeed, Tuan argues that one of the ways space is turned into place is through kinaesthetic familiarity – the habitual ability to move through it unthinkingly. To Tuan, though, the focus on experience is a holistic one. Alongside the habitual and pre-cognitive lie the processes of cognition that are able to reflect on and share ways of being in place. The importance of the visual is emphasized in Chapter Twelve (Visibility: the Creation of Place) in which he argues: 'Place can be defined in a variety of ways. Among them is this: place is whatever stable object captures our attention. As we look at a panoramic scene our eyes pause at points of interest. Each pause is time enough to create an image of place that looms large momentarily in our view' (Tuan, 1977: 161). Place, then, is a subtle mixture of both fairly self-aware reflection and the world of habitual action. We may know a place profoundly by participating in it in an unself-conscious way but we may also relate to places through the construction of something visible. This is how monuments, for instance, traditionally work. Place-making from the smallest corner of a favourite room to the nation and the globe happens both though practice and through a more symbolic register.

Tuan also pays considerable attention to the problem of time in *Space and Place*. Again this prefigures more recent assertions of the need to consider time alongside space in human geography (May and Thrift, 2001). Both space and place, he argues, have important temporal dimensions. The awareness of time and space comes together in the body. 'We have a sense of space because we can move and of time because, as biological beings, we undergo recurrent phases of

tension and ease' (Tuan, 1977: 118). Temporality is often described spatially. Thus we talk of things being forward or backward in time as well as space. Here is now. There is then. There is half an hour away. Observations such as these open up a plethora of diverse observations about how we humans inhabit time and space. Dancers, for instance, 'move forward, sideways, and even backward with ease. Music and dance free people from the demands of purposeful goal-directed life, allowing them to live briefly in what Erwin Straus calls "presentic" unoriented space' (1977, 128–129). Place too, Tuan tells us, has to be understood temporally. Place can be understood as a pause in time as well as space. A true sense of place takes time to establish. 'If we see the world as process' Tuan tells us 'we should not be able to develop any sense of place' (1977: 179). Mobility, in *Space and Place*, is part of the world of space and is antithetical to place and the kinds of attachment place necessitates. Finally, place can be a way of making time visible. We get to know a version of the past through the ways places are made to represent memories. This can include everything from the family photos we place on our mantelpiece to the statues of heroes that appear in city squares.

The division between space and place constructed by Tuan became a taken-for-granted distinction in human geography by the early 1980s. This distinction, more or less in the form Tuan suggested, is still taught to students across the English-speaking world. *Space and Place* is still widely used throughout the world, frequently cited as the foundational text of notions of place and admired across many disciplines and beyond the academic world entirely. My book, *In Place/Out of Place* (1996), would have been unthinkable without Tuan's inspiration. In that book I took the notion of place as a field of care and centre of meaning and developed it through a consideration of the exclusions that such attachments foster and what that might mean for people, practices and things that were deemed 'out of place'. This was typical of the way many younger geographers, during the 1990s, took the idea of place developed by Tuan and began to insert it into a world of cultural politics that Tuan mostly ignored. Benjamin Forest, for instance, revealed how place became an important symbol for new forms of gay identity in West Hollywood (1995) and Karen Till has shown how place is implicated in the construction and contestation of public memory in Berlin (2005). In each of these works Tuanian place was both adopted and adapted to consider issues of contestation and power as well as the more humanistic notions of meaning and experience.

Geographies of place, 30 years on from Tuan's book, share his concern for meaning, belonging and experience but also include reflections on power, exclusion and social difference that are largely missing from *Space and Place*. To some this has gone so far that the original interest in individual experience has been subsumed by wider social issues. A number of books have built on the humanistic perspective advocated by Tuan (Entrikin, 1991; Sack, 1997; Adams *et al.*, 2001). Sack, for instance, has built on Tuan's notion of place by considering the ways place as a centre of human meaning ties together worlds that are normally held apart – the worlds of nature, meaning and society. In addition, he has developed an ethical and moral framework for human action based on a Tuanian notion of place (1997, 2003). Tom Mels' edited collection, *Re-animating Place* (2004) attempts to recapture some of the humanistic insights of Tuan, along with Anne Buttimer, through an engagement with the rhythms of place as a grounded form of human experience. More importantly perhaps many of Tuan's ideas have travelled into geographies which are not explicitly humanistic and indeed to other disciplines entirely. The notion of place as a meaningful centre of human experience, for instance, has been developed by philosophers such as Casey (1998) and Malpas (1999).

Geographers and others, with backgrounds in Marxism, feminism and post-structuralism, have all developed critiques of Tuan's notion of place in particular (Rose, 1993; Massey, 1997; Harvey, 1993). Place, they argue, is far too vague a notion. It ignores power relations and forces of exclusion that work through a contested association of location, meaning and practice. Perhaps the most well-known critique of humanistic conceptions of place in general is Massey's development of the idea of a 'progressive sense of place' (also called a 'global' or 'extrovert' sense of place). Massey takes broadly humanistic discussions of place to task for insisting on a bounded sense of identity attached to place that is rooted in history in more or less linear ways (Massey, 1997). Places, she argues, are rather the product of multiple mobilities intersecting. These are open to change and associated with multiple, rather than singular, identities and histories. Places, as such, do not have clearly defined boundaries. Such a view is very different from Tuan's assertion that a world of process is antithetical to the construction of any kind of sense of place. Others prefer to use the notion of space in a more subtle way, suggesting a variety of figurative and literal spaces including social space, third space and lived space. Many have been inspired by the French urban theorist Henri Lefebvre, whose book *The Production of Space* (1991) introduced a notion of space which is socially produced, meaningful and lived. Abstract space, the kind of space discussed by Tuan, becomes just one version of space in Lefebvre. Lefebvre writes of 'representations of space' as the conceptual space of planners, politicians and social engineers. This closely mirrors Tuan's outline of space as a more abstract realm. But Lefebvre also writes of 'representational spaces' – 'space as directly lived through its associated images and symbols' (Lefebvre, 1991: 39). This is inhabited space and, as such, it is very close to Tuan's definition of place. Given the multiple definition of space in Lefebvre it is no surprise that place does not really feature in the work of Soja who, like Lefebvre, develops notions of space as both meaningful and practised, thus obviating the need for discussions of place (Soja, 1996).

Conclusion

As a way of doing geography *Space and Place*, like much of Tuan's work, is hard to pin down. One reason for this is that Tuan does not undertake 'research' in the usual sense of the word. He does not use any 'method' that is likely to appear in a methodology class. There is no evidence of ethnography, semiotics, statistical analysis or even in-depth archival work. The impression I get from reading Tuan's work is of someone who reads a lot of what is often called secondary material and makes connections between them that would otherwise have remained unnoticed. Many of Tuan's observations are assertions of common-sense.

Often they are supported by a range of anthropological literature, classic sociology texts, poetry, novels and snippets of psychology. It is almost as if Tuan hires an army of creative academics, writers and philosophers to do the research for him. The examples Tuan uses are as diverse as his sources but rarely include reflections on modern culture – particularly popular culture. We are more likely to encounter indigenous groups of Africa seen through the eyes of anthropologists or the plans of an ancient Chinese city than anything recognizably here and now. Tuan writes in a very accessible reflective way. It is, in many ways, a model of writing for a large educated audience that stretches far beyond the confines of a particular set of theoretical interests or, indeed, a particular discipline. It was this kind of writing that so engaged me back in 1985. A year later I was thinking about doing a PhD and was advised to contact the geographers whose work I most admired. I wrote to Yi-Fu and received a charming letter back. I went to Madison, Wisconsin to start my academic career.

Secondary sources and references

Abler, R., Adams, J. and Gould, P. (1971) *Spatial Organization: The Geographer's View of the World*. London: Prentice-Hall.

Adams, P., Hoelscher, S. and Till, K. (eds) (2001) *Textures of Place: Exploring Humanist Geographies*. Minneapolis: University of Minnesota Press.

Buttimer, A. and Seamon, D. (1980) *The Human Experience of Space and Place*. New York: St Martin's Press.

Casey, E. (1998) *The Fate of Place: A Philosophical History*. Berkeley, CA: University of California Press.

Cresswell, T. (1996) *In Place/Out of Place: Geography, Ideology and Transgression*. Minneapolis: University of Minnesota Press.

Entrikin, J.N. (1991) *The Betweenness of Place: Towards a Geography of Modernity*. Baltimore, MD: Johns Hopkins University Press.

Forest, B. (1995) 'West Hollywood as Symbol: The Significance of Space in the Construction of a Gay Identity', *Environment and Planning D: Society and Space* 13: 133–157.

Haggett, P. (1965) *Locational Analysis in Human Geography*. London: Edward Arnold.

Harvey, D. (1993) 'From space to place and back again', in J. Bird, B. Curtis, T. Putnam, G. Robertson and L. Tickner (eds) *Mapping the Futures*. London: Routledge, pp. 3–29.

Jackson, J.B. (1980) *The Necessity for Ruins*. Amherst: University of Massachusetts Press.

Lefebvre, H. (1991) *The Production of Space*. Oxford: Blackwell.

Ley, D. (1974) *The Black Inner City as Frontier Outpost*. Washington, DC: AAG.

Ley, D. and Samuels, M. (eds) (1978) *Humanistic Geography: Prospects and Problems*. London: Croom Helm.

Lorimer, H. (2005) 'Cultural geography: the busyness of being more-than-representational', *Progress in Human Geography* 29 (1): 83–94.

Malpas, J.E. (1999) *Place and Experience: A Philosophical Topography*. Cambridge: Cambridge University Press.

May, J. and Thrift, N. (eds) (2001) *Timespace: Geographies of Temporality*. London: Routledge.

Massey, D. (1997) 'A global sense of place', in T. Barnes and D.G. Gregory (eds) *Reading Human Geography*. London: Arnold, pp. 315–323.

Meinig, D. (ed.) (1979) *The Interpretation of Ordinary Landscapes*. Oxford: Oxford University Press.

Mels, T. (2004) *Re-Animating Place: A Geography of Rhythms*. London: Ashgate.

Relph, E. (1976) *Place and Placelessness*. London: Pion.

Rose, G. (1993) *Feminism and Geography: The Limits of Geographical Knowledge*. Cambridge: Polity.

Thrift, N. (2004) 'Summoning life', in P. Cloke, P. Crang and M. Goodwin (eds) *Envisioning Human Geographies*. London: Arnold, pp. 81–103.

Sack, R. (1997) *Homo Geographicus*. Baltimore, MD: Johns Hopkins University Press.

Sack, R. (2003) *A Geographical Guide to the Real and the Good*. New York: Routledge.

Soja, E. (1996) *Thirdspace: Expanding the Geographical Imagination*. Oxford: Blackwell.

Till, K. (2005) *The New Berlin: Memory, Politics, Place*. Minneapolis: University of Minnesota Press.

Tuan, Y.-F. (1974) 'Space and Place: Humanistic Perspective', *Progress in Geography* 6: 211–252.

Tuan, Y.-F. (1974) *Topophilia: A Study of Environmental Perceptions, Attitudes and Values*. Englewood, NJ: Prentice-Hall.

Tuan, Y.-F. (1977) *Space and Place: The Perspective of Experience*. Minneapolis: University of Minnesota Press.

8 THE LIMITS TO CAPITAL (1982): DAVID HARVEY

Noel Castree

The aim is ... to create frameworks of understanding, an elaborated conceptual apparatus,
with which to grasp the most significant
relationships at work within the intricate
dynamics of social transformation. (Harvey,
1982: 450–451)

Introduction

The Limits to Capital was first published in
1982. It is a book with epic ambitions. It aims
to explain how a raft of geographical phenomena – such as city-regions, nation states
and transportation networks within and
between them – are integral to the functioning of the world's dominant economic system
(capitalism). Harvey is arguably the most
famous contemporary geographer, known for
his politically engaged scholarship beyond, as
much as within, the field of professional geography. Author of over a dozen major works
(including *The Condition of Postmodernity*: see
Woodward and Jones, Chapter 15, this volume),
The Limits to Capital is Harvey's 'favourite text'
(2001: 10).

The book, which was nearly a decade in
the making, was written for two audiences.
As the leading Marxist geographer of his
day, Harvey hoped that *The Limits to Capital*
would influence research and teaching among
what, at the time, was a small group of like-
minded geographers (such as his former
students Neil Smith and Richard Walker). As
a geographical Marxist, Harvey also wrote

Limits to Capital in the hope of persuading a
fairly large community of academic Marxists
outside geography to take geographical questions more seriously than they had previously.
In short, *The Limits to Capital* was intended to
be a paradigmatic contribution: nothing less
than a modern geographical equivalent of
Karl Marx's magisterial work *Capital* (upon
which Harvey's book is in large measure
based). The one important contrast with
Marx's own writing is that Harvey's book was
largely an *academic* contribution, a significant
fact to which I will return later.

I will structure my account of *The Limits
to Capital* as follows. In the next section I will
set the book in its original context and
explain the significance of its status as a
largely academic work. I will then attempt to
summarize the book's arguments for the
benefit of student readers who, with few
exceptions, would not get far if they tried to
read *The Limits of Capital* unaided. In the
penultimate section I will consider the book's
impact within and beyond professional geography between 1982 and the date of its first
reissue (1999). My brief concluding section
relates *The Limits to Capital* to the immediate
circumstances of our time and likely future
scenarios for humanity worldwide.

Marxism, geography, capitalism

'Internalist' approaches to intellectual change
in academic disciplines focus on debates

within those disciplines. Such approaches presume that academia is relatively isolated from the wider society, and that intellectual progress emerges from the to-and-fro of criticism internal to the academic community. By contrast, 'externalist' approaches situate changing academic research in a wider societal context – a context seen to exert a more-or-less direct influence on what academics choose to do. *The Limits to Capital* must be understood in *both* its disciplinary and wider societal contexts, as I will now explain, starting with the former.

When he wrote *The Limits to Capital*, Harvey was a Professor of Geography at the Johns Hopkins University, in the east coast US city of Baltimore. Prior to this he had been a Lecturer in Geography at Bristol University, England. In Bristol, Harvey was exposed to the so-called 'new Geography' being pioneered by the likes of Richard Chorley (at Cambridge University) and Peter Haggett (a Bristol University professor). In their view, Geography should become a 'spatial science': one whose special role within the family of sciences was to rigorously describe and explain the location of phenomena on the earth's surface. Inspired by the 'new Geography', Harvey wrote the definitive geography book on scientific method (*Explanation in Geography,* 1969; see Johnston, Chapter 4, this volume). This book gave further impetus to spatial scientists' preoccupation with hypothesis testing, the search for spatial regularities and laws, the development of general models and theories, and the use of statistical techniques.

However, by 1969 a decade of scientific geography was already proving too much for Harvey. The reason was a gap between what was going on inside the discipline of geography and what was going on in the wider world. The disciplinary and societal contexts collided, and Harvey wanted to realign them by transforming professional geography from within. The late 60s and early 70s were turbulent years. The post-war economic boom in the West came to an end with crippling oil price hikes; the US civil rights movement was in full flow; the modern environmental movement came into existence; the 'imperialist' war in Vietnam rumbled on until the US was obliged to withdraw its troops; dissidents struggled against French and British colonial powers in Africa and beyond; there were the famous student riots and strikes in Paris in May 1968; feminism made waves throughout the so-called 'developed' world; and trade unions were at loggerheads with national governments across the Western world. The solidities of the post-1945 era seemed suddenly to be melting into air. As Harvey (2001: 5–6) noted retrospectively, 'I turned in my *magnum opus* to the publishers in May 1968, only to find myself acutely embarrassed by the change in political temperature … I realised I had to rethink a lot of things I had taken for granted …'.

In *Social Justice and the City* and a string of now classic essays, Harvey (1973) became known as an uncompromising critic of spatial science and a pioneer of Marxist geography. He called for nothing less than a 'revolution in geographic thought' in which a Marxian research agenda would replace the spatial science paradigm. In Harvey's view, this paradigm was profoundly problematic. In essence, its claims to 'objectivity' and 'neutrality' concealed its *complicity* with a profoundly unjust social order that operated in the interests of power elites nationally and globally. 'Scientific' studies by geographers of migration patterns in space and time, of traffic congestion, or of how to manage water resources dealt with symptoms not causes in Harvey's estimation. They remained blind to the true nature and origins of myriad economic, social and environmental problems. What is more, for Harvey spatial science wrongly presumed that academic research could and should be impartial when, in fact, it was nothing more than a *non-neutral part* of the reality it sought to comprehend. In Harvey's view, university research and teaching did not merely 're-present' truths existing 'out there' in the wider world. Instead, they helped to *constitute* that wider world insofar as academic 'expertise' was used to shape student, governmental, business and public behaviour.

Marx was a famous – for some notorious – nineteenth-century philosopher, political economist, journalist, pamphleteer and labour movement leader. Dismayed by the appalling conditions in which growing numbers of working-class people were forced to live and work, he sought to understand how and why so few people get rich while so many remain poor. In his later writings, such as *Capital* volume one, Marx argued that *capitalism* was a major cause of social inequality (and environmental degradation) in the Victorian world. The appeal of Marx's work to Harvey was arguably two-fold. First, Marx's focus on capitalism offered a way to talk about deep causes rather than the surface problems – the kind of problems that so many spatial scientists studied and sought to ameliorate through public policy prescriptions. Secondly, Marx's later work took an 'activist' view of knowledge in which to *study* the world is to *change* the world, given the right conditions. Marx saw 'knowledge' as the polite world for competing *ideologies* that serve specific societal interests. Marxism was intended to be the ideology of the working class: a set of explanatory tools that can help this class realize its own true interests by creating a more egalitarian mode of production (a socialist one).

In *Social Justice* and subsequent essays published in the 1970s, Harvey did indeed precipitate something like the revolution in geographic thought he wished for. His work especially appealed to a younger generation of geographers who had been politicized by the turbulence of the post-1968 period. But ironically, Harvey's radicalization of geography (and more particularly the human side, since physical geographers carried on much as before) did not have any wider societal impacts. In part this was because Marxist geography fast became an *academic* movement, one that conformed to the professional standards of Western academia. Even Harvey, socialized as he was into the norms of elite universities, tended to focus his writing activities on high-level, difficult works like *The Limits to Capital* – works unlikely to attract a wide readership. Secondly, even if the early Marxist geographers like Harvey *had* connected their work more vigorously to workers' movements, the prospects for achieving a more socialist society had diminished by the late 70s.

The optimism of the late 60s gave way, in less than 10 years, to despair within and beyond the West. More authoritarian governments (e.g. those of Margaret Thatcher) would not tolerate worker demands for better pay and conditions; meanwhile, global economic restructuring led to job losses in many highly unionized countries and job gains in areas with little or no history of labour organizing. In short, by the time Harvey handed over the manuscript of *The Limits to Capital* to his publisher, academic geography had been radicalized in a left-wing direction but the wider world seemed more conservative than it had been when Harvey made his initial turn to Marxism a decade earlier.

The Limits to Capital explained

The Limits to Capital is a text about the horrors and glories of the capitalist world in which we live. Given how insinuated into most aspects of people's lives capitalism now is, Harvey's book has a universality of relevance that most books by academics cannot equal. To appreciate the distinctiveness of the text we need to start with some comments about theory, about the kind of 'geography' that has long interested David Harvey, and about something called 'dialectics'.

The Limits to Capital is a work of theory. In simple terms, to theorize is to engage in a process of abstract, rather than concrete, argumentation. To 'abstract' means to 'take from': abstraction thus involves pulling something out of its context of operation in order to understand how it works. Theorists thus do mentally what clinical researchers do surgically. They dissect a whole (e.g. a social system) in order to inspect the parts so that the particular function of these parts and their inter-relationships can be understood. Generally speaking, and to use cartographic imagery, 'theorists' therefore try to represent social (and/or biophysical) reality in terms of a conceptual map that highlights the key topographical features, their relative positioning and their relationships.

So much for theory. What about geography? There are three things to say here. The first is that Marx and most of his intellectual successors did not pay much attention to geographical issues. This meant that most Marxists understood capitalism in non-geographical or 'aspatial' ways: as if it existed on the head of a pin without the need for material landscapes of production, transportation, consumption and social reproduction. Secondly, this inattention to geography was linked to a false equation between the sorts of things that geographers study and the realm of 'facts'. For decades, non-geographers (and even a few geographers!) have assumed that geography cannot be a theoretical discipline because the sorts of things it studies are 'contingent'. A contingent occurrence – like the construction of a new airport or the incidence of famine – will almost always have causes that can be represented theoretically. For instance, a relationship of gender inequality between men and women can be described and explained conceptually (as an outcome of 'patriarchal' social relations). But this relationship will manifest itself in, and produce, different contingent forms – for instance, particular heterosexual arrangements in the home,

specific workplace dynamics and so on. Accordingly, the contingent forms are connected to but do not in themselves constitute the 'underlying' processes generating them. Contingencies must be studied *empirically*, and because they often *vary* in time and space geographers are chief among those who investigate them.

Harvey dissents from this view. Since *Social Justice* he has argued that geographical phenomena can be theorized. This is the same as saying that these phenomena have a *constitutive* role to play in the fundamental processes that give rise to them in the first place. Here, then, there is no distinction between 'process' and 'outcome' because the latter makes the former flesh and, once it exists, may affect the subsequent operations of the process in question. This brings me to a third and final observation about the kind of 'geography' that interests Harvey. Going back to his doctoral thesis on the Kent hop industry, Harvey had long been fascinated with geographical difference: with the rich *specificities* of people and place. However, as Harvey shows in *The Limits to Capital* and elsewhere, we can make some *general* observations about these otherwise unique geographies. More specifically, Harvey argues that there are some 'signature geographies' – or characteristic patterns – associated with capitalism's past, present and (so long as it exists) future. These include the growth and decline of city-regions housed within nation states and the construction of elaborate transportation networks to more closely connect them.

Lastly, a word about dialectics. Dialectics is both an analytical tool (an epistemology and method) and a broad description of how social and biophysical systems work (an ontology). In both cases *contradiction* is a key idea. In the ontological sense, a dialectical approach says that some social and biophysical systems contain opposing tendencies, the collision of which leads to change in (and even the dissolution of) those systems. In the conceptual and methodological sense, a dialectical approach obliges the theorist to apply their mental scalpel in a specific way. Faced with a large complex system (like capitalism), the theorist makes a series of mental 'cuts' that reveal one set of processes; subsequent cuts reveal other processes that can then be shown to both arise from and yet contradict those identified earlier in the analysis. This means that when one reads the work of a dialectical thinker like Harvey the *beginning* of the analysis only really makes sense at the *end* of the book or essay in question.

Marx's original argument

Drawing upon *Capital* and other late works by Marx, Harvey theorizes capitalism as a process rather than a thing. This is entirely appropriate because the capitalist mode of production is really about *circulation*. As Marx argued, capitalism can be described as shown in Figure 8.1.

Here owners of the means of production (capitalists) advance a sum of money to purchase three things: (i) labour power (LP: people who must sell their skills and time for a wage in order to live), (ii) inputs (e.g. parts, raw materials), and (iii) infrastructure (e.g. premises, machines) [ii and iii = MP]. These things are then deployed in the production (P) of new commodities (C★) which are sold in order to recoup the original sum of money advanced plus an increment.

In essence, capitalism has one over-riding logic: *profit*. It is thus a system in motion because capitalists are constantly aiming to make returns on their investments during each round of commodity production. What Marx called 'capital' was precisely this process of ceaseless accumulation which assumes different physical forms between the beginning and end of each production cycle. As part of this process, firms are compelled (unless they

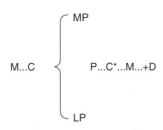

Figure 8.1 The primary circuit of capital accumulation

enjoy monopolies) to *innovate* technically and organizationally. This is because they must *compete* for market share with rival firms in their sector. Clearly, this has implications for waged workers because their employers will seek to alter salaries and benefits, hours of work, number of employees, and much more besides. In sum, capitalists as a class engage in a constant *struggle* with workers to extract profits because, for Marx, workers have the special quality of producing more economic value than they themselves are normally paid to do their jobs. This struggle produces winners and losers at any one time. Some employees fare well and others do not; some firms go under for lack of 'competitiveness', while others prosper by laying-off workers and substituting them with 'smart machinery'.

If, then, accumulation, inter-firm competition, innovation and class struggle are all interconnected elements of capitalism it is in essence a *contradictory* mode of production. This makes it a *crisis-prone* system precisely because its central features are profoundly inconsistent. For instance, as wage workers capitalists see people as a cost of production to be minimized and made more efficient. But as consumers capitalists see people as spenders of money without whom they cannot realize profits. Here, then, is one of several contradictions: in this case between the points of production and the points of commodity sale, because most people fulfil a *double* function in capitalist societies as earners and spenders of money. Marx and his acolytes hoped that when 'over-accumulation crises' occurred – as Marx thought they inevitably would from time-to-time – the working class as a whole would rebel against the capitalist system. In Western countries this has never come to pass, though it did occur in Russia in 1917 led by Vladimir Lenin (founder of the now defunct Soviet Union). This raises the important question of how capitalism's contradictions are managed, by whom, and with what specific consequences.

Harvey's contribution

The title of *The Limits to Capital* says much about the book's contents, as one might expect. The 'limits' referred to by Harvey are two-fold. The first are those of Marx's original arguments, especially their inattention to geographical issues. The second are the limits of capitalism as an economic system. *The Limits to Capital* seeks both to extend Marx's theory of how capitalism works and, in so doing, to more comprehensively identify capitalism's internal contradictions. As a purely theoretical book, *The Limits to Capital* is intended to specify the 'basic laws of motion' of capitalism past, present and future. In other words, it deliberately does not offer a close empirical demonstration of how capitalism works 'on the ground' in real times and places.

The first seven chapters rehearse the arguments sketched in the previous sub-section. Harvey recapitulates Marx's propositions in some conceptual detail, leading to what he calls 'the first cut theory of crisis'. However, it is the six chapters of the second half of *The Limits* that capture the book's true originality. As with the early chapters, crisis is the key organizing them. The crisis tendencies Marx identified clearly pose a major problem for the capitalist mode of production. Potentially, they can precipitate social revolutions far and wide. Even when they do not, these tendencies can pitch whole societies into chaos – with even the state badly affected as it tries bravely to manage the problems for both capitalists and workers alike (remember that the state, as an institution, depends upon revenues generated within a healthy capitalist economy [in the form of taxes, for example]). The question then arises: how can crises of over-accumulation be avoided? Harvey addresses this question, in the first instance, with reference to the financial system.

Finance capital is, in basic terms, credit: it is money lent with a view to receiving interest from the lender. Though lending institutions existed long before capitalism ever did, they have a pivotal role to play in capitalist societies. This is so in two senses. First, a portion of the surplus generated in the 'primary circuit of capital' (Figure 8.1, above) can go into short- or long-term savings by firms (and workers too save money, of course). Banks, building societies and other savings institutions can then lend this money to new firms or to state bodies in order to finance particular investments. Secondly, a portion of the surplus generated in the primary circuit can be diverted directly to investment. For instance, successful firms in the US can buy stocks-and-shares in new business ventures in 'emerging economies' like Bulgaria. In sum, financial institutions oil the wheels of capitalist commerce because they divert large amounts of capital from the primary circuit into future opportunities for commodity production, transportation and consumption. How, then, does this relate to the possible avoidance of economic crisis? On the one side, the financial circuit of capital provides what Harvey calls a 'temporal fix'. On the other side, it is central to what Harvey calls a 'spatial fix'. Let me take each in turn even though, in reality, they are two sides of the same coin. This requires a brief discussion of the role of the built environment.

For the primary circuit of capital to function successfully a portion of it must be composed of four kinds of built environments: namely, those of production (e.g. factories), consumption (e.g. shopping malls), reproduction (i.e. homes) and distribution/communication (e.g. airports, rail systems). These built environments are, if you like, the arteries that make the process of capital accumulation possible. There are two unique things about them. First, they are extremely expensive to construct (for instance, a new international airport can easily cost over one billion US dollars). Secondly, these costs mean that few lenders acting alone can afford to finance them. This is where the financial system as a whole comes to the rescue. By pooling the revenues of countless firms and investors, financial institutions (and, for that matter, many national governments) can 'switch' large amounts of capital out of immediate production to pay for large-scale and long-lasting infrastructural investments. Given that the financial returns for investors are potentially uncertain and very long-term, such institutions use fictitious capital to incentivize these investors. As Harvey reminds readers in *The Limits to Capital*, fictitious capital is in effect a set of promissory notes that make claims on a future share of profits from the primary circuit.

What all this means, as Harvey shows in Chapters Eight, Nine and Ten of *The Limits*, is that the financial circuit of capital is at once autonomous from and yet tethered to the primary circuit. In facilitating capital switching out of immediate commodity production, financial institutions can enable the much-needed production of the four types of built environment listed above. In what ways does this switching constitute a 'temporal fix' for capitalism's crisis tendencies? Harvey argues that tying-up surplus capital (and labour) in the construction, use and maintenance of new built environments is a way of *deferring* crisis. It postpones for many years (potentially at least) the proverbial chickens

coming home to roost. This is because things like road networks and factories only realize profits (if they do at all) in the medium-to-long term, but rarely ever in the short-term. However, none of this eliminates crises tendencies in the capitalist system. Harvey's 'second cut' crisis theory suggests that, eventually, new investments made via the financial system will fail to produce the wealth necessary to pay interest on those investments. At that point, a long-deferred crisis of more-or-less general proportions could engulf the capitalist world.

Let us now turn to the 'spatial fix', which is central to the closing chapters of *The Limits to Capital*. For commodity production to occur, it is obviously necessary in most cases that workers and employers be proximate. It is equally necessary that both parties have built environments of the four kinds mentioned in order to undertake necessary activities locally and to connect with the wider capitalist world. Over time, a particular economic activity, or ensemble of activities, comes to characterize a particular town, city or city-region and its inhabitants. These places, Harvey argues, have a 'structured coherence' or internal integrity. Cross-class alliances can develop to protect local interests and investments against competing places in the global economy. However, according to Harvey there is a problem. Given capitalism's chronic crisis tendencies, it frequently becomes functional to switch large amounts of capital out of 'developed' towns and city-regions and into less developed ones or ones that have fallen on hard times. This switching can involve investing in new built environments abroad in the hope that they support new productive activities that will, in time, pay interest on the original investments. This constitutes a 'spatial fix' for the obvious reason that geographical expansion is here being used to find an outlet for excess capital.

Harvey's ultimate conclusion is that spatial fixes (like temporal ones) can never prevent capitalist economic crises, only defer their ultimate occurrence. Indeed, he argues that they simply *widen* the landscape across which capitalism's inner contradictions operate. It is precisely in allowing these growth poles to emerge that formerly profitable places suddenly face geographical competition – which can produce local, regional and national crises as former 'winners' become present-day 'losers'. In short, *uneven geographical development* is an intrinsic, non-accidental part of capitalist life. Given this contradictory logic, Harvey ends *The Limits to Capital* on an apocalyptic note – his 'third cut' theory of crisis. Eventually, he argues, spatial and temporal fixes together will no longer provide a safety valve for the system as a whole. At that point, there is likely to be a mad scramble (as he argued happened in the late 1930s) among capitalist countries to avoid suffering the worst consequences of economic crisis. (For a summary account of *The Limits to Capital* see Harvey, 1985.)

Evaluating *The Limits to Capital*: from the first to the second edition

Needless to say, the previous section barely does justice to the arguments of *The Limits to Capital*. How then does one evaluate a book like *The Limits*? The answer is by no means obvious. *The Limits to Capital* is widely known within and beyond geography but not, in my view, widely read, despite a reissue in 1999. Even though the book is highly cited by left-wing social scientists, this does not mean they have read the book in detail, let alone understood its intricate arguments. Does this mean that the book should be regarded as 'a failure' – undeserving of a chapter-length treatment in a book on key texts? My own view is that *The Limits* is a towering achievement. The sheer difficulty of understanding the book is not, I submit, a negative sign of its quality or its importance. I say this for the following reasons.

First, one needs to look at who precisely *has* read, understood and been influenced by *The Limits of Capital*, rather than worry about who has not. As I noted in the introduction, Harvey wanted *The Limits to Capital* to be a foundational text for both Marxist geography and geographical Marxism. In that he has surely succeeded. In the years between the book's original publication and its reissue, Marxist geography became not only the dominant left-wing approach in Anglophone human geography, it also helped make left-wing geography as such a central element of Anglophone geography *tout court*. This was no mere coincidence. *The Limits to Capital* specified a new research agenda of such profundity that it attracted some of the best minds in human geography throughout the 1980s. Without it, key Marxist books like *The Golden Age Illusion* by David Rigby and Michael Webber (1996) could not have been written. Likewise, without the credibility that *The Limits to Capital* lent Marxist geography at large, people like Neil Smith, Don Mitchell, Kevin Cox and Richard Walker could not have gone on to assume important positions in world-leading geography departments. Outside geography, albeit belatedly, key Marxists like Fredric Jameson and the sociologist Bob Jessop began to read Harvey's *magnum opus* closely, once the best-selling *The Condition of Postmodernity* had alerted them to his prodigious talents. That a special issue of the left-wing academic journal *Antipode* was published to mark *The Limits to Capital*'s twentieth anniversary says much about the book's importance to Marxists within and outside academic geography (Brenner *et al.*, 2004).

Secondly, a book's significance can also be measured by who its critics are as much as its fans. Such is the range of topics covered in Harvey's book and such is the depth of treatment each one receives, that it has become a touchstone for leading human geographers of a post- or non-Marxist persuasion. A good example of this is *Money/Space* by the British geographers Andrew Leyshon and Nigel Thrift (1997). This book was a formative contribution to the geographical study of money in the modern world. It lays out a broad research agenda for the interrogation of where, why and in what specific forms money moves, rests and moves again. Even though the book's authors were not Marxists, their text took Harvey's arguments in *The Limits to Capital* as a key point of departure (see Leyshon, 2004). For them, Harvey's understanding of money missed, among other things, its cultural importance by emphasizing its economic functions. This is one of several topic areas where human geographers have developed new lines of inquiry through a critical engagement with Harvey's strongly Marxist position.

These two positive points having been made, one still cannot escape the fact that outside academia Harvey's book is almost totally unknown. This would seem to be an indictment of Harvey's own professed desire to change the world by altering people's understanding of their daily lives. However, in Harvey's defence, *The Limits* was not intended to be read by a wider, non-academic audience. Why, then, did he write the book in the way he did? A cynical response to this question might suggest that Harvey was slavishly playing the academic game. In other words, Marxist or not, he was simply following the well-worn path of previous successful academics: write a 'major book' in order to gain academic credibility. However, this is surely too cynical a view. After all, Harvey's academic reputation was *already* secure by the late 1970s, at least within geography. This suggests an alternative explanation for the book's density. Quite simply, capitalism *itself* is so complicated that only a book of equal complexity could do justice to its various (il)logics. Having written the book, it was then up to Harvey himself and other Marxist geographers to translate its insights into more intelligible forms for

workers' groups, trade unions and the like. That such acts of translation never occurred on a large scale is no indictment of *The Limits to Capital*. Instead, it says much about the prevailing political mood in the West through the 1980s and 90s. Outside academia Marxism became something of a dirty word during these years.

Conclusion

It may seem strange that *The Limits to Capital* was reissued in 1999. In the West, and more generally, the last 25 years have been ones when conservative and neo-liberal agendas have dominated politics, business and public debate. The 'death of communism' in the late 1980s (when the former USSR and the Eastern Bloc dissolved) seemed to indicate the 'superiority' of capitalism over socialism. In this context, Marxist theory was not only 'off-message' in the wider society; more pointedly, it seemed to have had its arguments confounded by capitalism's remarkable ability to prevent economic crises becoming anti-capitalist revolutions. Additionally, from the early 1990s Marxism lost its previous academic dominance within and outside Anglophone human geography. A new generation of academics chose to make their names on the backs of non-Marxist theorists like Jacques Derrida, Michel Foucault, Bruno Latour, Paul Virilio and many others besides. By 1999, Harvey was seen by many younger left-wing academics as a bit of a 'dinosaur' – an old-fashioned 'meta-theorist' who was inattentive to the complexity and contingency of social life. Given these facts, reprinting *The Limits to Capital* may have seemed an act of folly.

I say *seemed* deliberately. When leading left-wing publisher Verso chose to introduce *The Limits to Capital* to a new generation of academics and activists it did so for good reason. Since the late 1990s, something of the revolutionary fervour of the late 60s/early 70s has reappeared in public life worldwide. The often-violent protests against the World Trade Organization, the Make Poverty History campaigns, the armed struggle of the Zapatistas in southern Mexico are three of many possible examples of present-day discontent with the manifest injustices of life in a capitalist world. In this new context the book arguably has a greater relevance than at any time since it was published. Nonetheless, it will never be an easy book to read and digest. If Marxists like Harvey are really to change the world for the better, there needs to be more engagement of the public and activists than heretofore (see Castree, 2006). This will be no easy task given how closeted in the 'ivory tower' most present-day academics are – even political academics like Harvey himself. Meanwhile, within academia itself, there are precious few Marxists of my own generation who can continue the tradition of spatialized Marxism that Harvey helped to inaugurate. In the years ahead, we may well confront the paradox that Marxism has few expositors left in the university world, while a large audience receptive to Marxist ideas exists out there in the wider world.

Secondary sources and references

Brenner, N., Castree, N. and Essletzbichler, J. (eds) (2004) 'David Harvey's *The Limits to Capital* two decades on', *Antipode* 36 (3): 401–549.

Castree, N. (2006) 'Geography's new public intellectuals?', *Antipode* 38 (2): 396–412.

Harvey, D. (1969) *Explanation in Geography*. London: Arnold.

Harvey, D. (1973) *Social Justice and the City*. London: Arnold. Reissued in 1988. Oxford: Blackwell.

Harvey, D. (1982) *The Limits to Capital* (Oxford: Blackwell). Reissued in 1999 and 2007 (London: Verso).

Harvey, D. (1985) 'The geopolitics of capitalism', in D. Gregory and J. Urry (eds) *Social Relations and Spatial Structures*. London: Macmillan, pp. 163–185.

Harvey, D. (2001) *Spaces of Capital*. Edinburgh: Edinburgh University Press.

Leyshon, A. (2004) '*The Limits to Capital* and geographies of money', *Antipode* 36 (3): 461–469.

Leyshon, A. and Thrift, N. (1997) *Money/Space*. London: Routledge.

Rigby, D. and Webber, M. (1996) *The Golden Age Illusion*. New York: Guilford.

UNEVEN DEVELOPMENT (1984): NEIL SMITH

Martin Phillips

At the very least, uneven development is the geographic expression of the contradictions of capital ... The historic mission of capital is the development of the forces of production via which the geographical equalization of conditions and levels of production become possible. The production of nature is the basic condition for this equalization, but equalization is continually frustrated by the differentiation of geographic space. (Smith, 1984: 152)

Introduction

In introducing Neil Smith's (1984) *Uneven Development*, I will begin with something of an admission. As a Master's student with an interest in Marxist ideas of political economy and uneven development, I bought this book soon after its publication in 1984 but struggled to really 'get into' the book, preferring instead David Harvey's (1982) *The Limits to Capital* for its steady and accumulating lines of argument (see Castree, Chapter 8 this volume). However, I returned to Neil Smith's book in later years, particularly when preparing to write a book on society and nature (Phillips and Mighall, 2000), at a time when I had become very interested in other writings by Neil Smith on gentrification (e.g. Smith, 1996). Re-reading the book after so many years, I began to think about the different ways in which I engaged with it, and whether this might say something about the book itself or simply says something about my changing interests and ability to tackle

what was, for me at least, initially quite a difficult book. Though I do not wish to discount the latter, I am going to suggest that my experiences reveal something significant about the character of the text and the varied interests of its author, arguing that there are connections between my reading of the book and its wider reception over the years.

The text and its author

As a reader I initially approached *Uneven Development* with two quite different interests in mind, namely political economy and society–nature relations. In re-reading the book it was re-assuring to find that these two concerns appear to have been those of its author as well. In its preface, for instance, Neil Smith states that the book 'represents the meeting of two types of intellectual investigation', namely a concern with 'renovating the terribly archaic conception of nature that dominates western thought' and an examination of the process of uneven development which emerged out of being 'fascinated with the process of gentrification' and a belief that this 'was itself a product of spatially more universal, if quite specific, forces operating at different scales' (Smith, 1984: vii).

Smith argues that his book seeks to integrate these two concerns, suggesting they have 'until recently enjoyed little serious cross fertilisation' (Smith, 1984: ix). Accordingly, the book might be read as two books within one, with Chapters 1 and 2 being concerned about nature and introducing the conception of socially produced nature, while the majority of the book (Chapters 3 to 5) focuses on the spatiality of economic development and the concept of the social production of space. However, such a bifurcated reading misses a clear line of argument running through the book that acts to link the two sections: the idea that nature and space are socially produced. The argument, clearly outlined in the book's introduction, is that the latter is very much predicated on the former, and that misunderstandings of the former have hindered understandings not only of nature but also of space.

Smith begins *Uneven Development* by highlighting the complex and contradictory ways in which nature has been defined, before arguing that these views enact one of two conceptualizations, namely either depicting nature as *that which is external*, a 'realm of extra human objects and processes existing outside society' (Smith, 1984: 2) or portraying nature as *that which is universal* (i.e. nature as a characteristic or inherent state). Smith suggests this dualism can be seen in the philosophical writings of Kant, although he argues it has both earlier roots and more recent manifestations. With respect to the latter, he focuses on two particular 'modes of expressing and conceptualising nature' (Smith, 1984: 3), which he identifies as 'the scientific' and 'the poetic'.

In discussing science and nature, Smith focuses particularly on the work of Bacon and Newton, suggesting they both utilized and promulgated concepts of external and universal nature. The latter sense of nature is implied, for instance, in notions of a 'scientific method', which Smith argues 'dictates an absolute abstraction both from the social context of the objects under scrutiny and from the social context of the scientific activity itself' (Smith, 1984: 4). By way of illustration, Smith comments on Newton's 'explanation' of an apple falling due to gravity:

> When he watched the apple fall, Newton did not ask about the social forces and events that led to the planting of the apple trees ... Nor did he ask about the domestication of fruit trees that gave the apple its form. He asked, rather, about the 'natural' events defined in abstraction from its social context. (Smith, 1984: 4)

With reference to the poetic, Smith focuses on the concept of wilderness as expressed in North America by landscape artists such as Cole, Church and Durant and writers such as Emerson and Thoreau. Smith suggests that the poetic concept of nature presented in such works, although appearing quite different from, and at times quite antagonistic to, the scientific perspectives, still relied on the twin dualism of external and universal nature. In particular, Smith emphasizes that the 'wilderness', from the period of nineteenth-century romanticism onwards, has been seen both as a place to escape from society and as a place where people can attain some transcendental/ universal union with nature.

Smith argued that this poetic vision of nature is ideological, which he defines as a viewpoint that is 'an inverted, truncated, disturbed reflection of reality ... rooted in ... the practical experience of a given social class' (Smith, 1984: 15). In particular he suggests that the universal concept of nature is used to 'invest certain behaviours with the status of natural events' whereby they are seen to be universal in both space and time: a norm to be enacted everywhere and unchanging throughout the past and into the future. For Smith both elements of the dualistic concept of nature are, however, bourgeois constructs, derived from and serving to justify capitalistic interests.

Having outlined the dualistic concept of nature, Smith turns his attention to undermining it by highlighting how nature is *socially produced*. For Smith the work of Karl Marx provides key insights for the development of a non-ideological concept of nature, even though he admits this concept is not 'laid out completely or succinctly' (Smith, 1984: 34) in the writing of this nineteenth-century self-styled 'political economist'. Nonetheless, Smith suggests that a non-dualistic account of nature can be discerned in Marx's writing, one that is both logical and historical in character. Here Smith emphasizes the concepts of *abstract* and *concrete* theory mentioned by Marx (subsequently elaborated in discussions of realism by the likes of Sayer, 1984), suggesting that nature should be seen as socially produced both in a general or abstract sense, and also more specifically, or concretely, as through time it becomes increasingly produced 'for exchange' and 'by capitalist production' (see Table 9.1).

Smith argues that in the most abstract sense, nature can be seen as socially produced in the dual sense that people, in order to live, have to work on nature and, in the process, are affected by nature. Smith suggests that transformation of nature through work is a 'natural activity' in the sense of being 'an eternal nature-imposed necessity' (Smith, 1984: 35). In other words, production is a natural activity in the universal sense that human life would not be possible without. Furthermore, the activity can be seen to involve nature as 'external' in that, to use the

highly gendered language in which Marx wrote, 'man ... opposes himself to nature ... setting in motion arms and legs, head and hand, the natural forces of his own body, in order to appropriate nature's production in a form adapted to his own wants' (Marx, 1976: 177; quoted in Smith, 1984: 36). However, within this quote one can see Marx both employing the notion of external nature and also subverting it through reference to a universal construction in which humans are placed *in* nature. Furthermore, as Smith highlights, Marx also subverted universal notions of nature by suggesting that in the act of appropriating nature through production, humans not only change external nature but also their own nature, changing their physiology through the development of tools, their material needs, their consciousness and their social relationships with each other. Society and nature are, Smith argues, brought into unity through the natural activity of human labour on nature, which in turn creates, or produces, new forms of human and non-human nature.

For Smith, the social production of nature is centred around what Marx termed '*use-value*'

Table 9.1   Smith's logico-historical construction of the production of nature

Level of abstraction	Production of nature characteristics
Production in general (Most abstract)	Production is a natural activity in the universalistic sense in that without it there would be no life. Production involves the appropriation of external nature utilizing the natural capacities of humans but in engaging in this activity both human and non-human nature is transformed.
Production for exchange	Productive activity is orientated towards market exchange and there is extension in the scale of the production of nature that involves a movement towards second nature.
Capitalist production (Most historically/spatially specific)	Production occurs through the wage–labour relationship in which people who have become separated from the means of production (i.e. land, equipment, resources, finance) sell their labour power to capitalists who own the means of production and who extract a profit by paying labour less than it produces in value. Capital requires continual production to extract profit and there is hence an inherent drive to consume nature. First nature becomes transformed into second nature.

in that human labour on nature is focused on satisfying basic, natural, needs such as sustenance and shelter. He also adds that whilst this abstract concept of the social production of nature offers some important insights, particularly in relation to challenging some of the precepts underpinning the dualistic constructions of nature previously identified, it is still of 'fairly limited' value (Smith, 1984: 38). What is needed, he suggests, is to examine more 'concrete', historically specific forms of human activity and the production of nature. In particular, Smith draws attention to the emergence of 'production for exchange' and 'capitalist production'. The former relates to the emergence of market exchange societies whereby production is determined not by the use-value a particular product may have but more by the amount of other products, or money, that might be exchanged for it. Smith argues that this historically emergent form of production represents an extension of the scale of the production of nature such that '[h]uman beings produce not only the immediate nature of their existence, but produce the entire societal nature of their existence' (Smith, 1984: 44). As with production in general, Smith discusses societal nature in terms of object of production, human characteristics, reproduction processes and human consciousness. He suggests that, amongst other things, one can see the use of non-human nature being determined by exchange as opposed to use value, growing separation between producers and the products of their labour, deepening gender divisions in labour, and an increased differentiation of mental activity from manual labour, such that nature itself becomes the subject of mental contemplation as well as productive labour. Smith (1984: 45) suggests that it is within this mode of production that a 'cleavage is created between nature and society' – a division which he elaborates in relation to the emergence of '*second nature*'.

Smith argues that the concept of second nature, which he traces back to the work of Greek and Roman philosophers such as Plato and Cicero, both emerged with, and sheds light on, exchange economies. Smith uses the terms 'first' and 'second' nature to refer to pristine and humanly modified nature respectively. As such, a wild landscape, plant or animal might be described as an element of first nature, while a domesticated landscape – an agrarian landscape of cultivated fields, a cultivated garden plant or a carefully bred domesticated animal – might be described as instances of second nature in that they are the product of quite intensive human labour although still widely viewed as being natural because their existence is dependent on non-human forces and processes.

This notion of first and second nature has been widely employed in some more recent discussions of society and nature relations (e.g. see Castree and Braun, 2001). Smith, however, utilizes the terms in at least two further senses that differentiate and intertwine with each other in a manner akin to a triple helix (see Figure 9.1). First, he uses such terms to refer to non-human and human entities respectively, or what might conventionally be described as 'the natural' and 'the human'. Such a terminology, however, effectively enacts the nature as external dualism of which Smith is critical, and by utilizing the notion of first and second nature, Smith is able to make reference to human/non-human differences without implying that the human is non-natural and also to highlight how notions of nature as universal come to be applied to human relations, such as the commodity relations of market society. Finally, Smith uses the terminology of first and second nature to refer to a distinction between materiality and abstraction that he sees as central to understanding capitalist production. He argues, for instance, that while within production for exchange, 'the difference between first and second nature is simply the difference between the non-human and the humanly created worlds' (the second sense of first and second natures discussed above), within

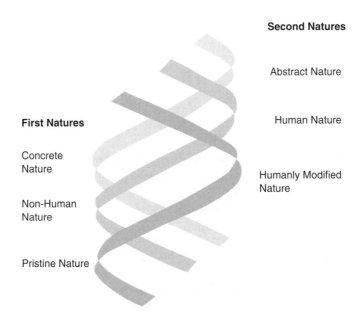

Second Natures

Abstract Nature

First Natures   Human Nature

Concrete
Nature   Humanly Modified
                               Nature
Non-Human
Nature

Pristine Nature

Figure 9.1   Senses of first and second nature in uneven development

capitalist production 'the distinction now is between a first nature that is concrete and material, the nature of use values in general, and a second nature which is abstract, and derivative of the abstraction from use-value which is inherent in exchange-value' (Smith, 1984: 55).

Smith then outlines a classic Marxist account of capitalist production in which labour has become divorced from the means of production and hence has to sell itself to capitalists who extract surplus value (or profit) by paying the workers less in value than they produce. There is therefore a continual drive to produce, for without this there can be no extraction of surplus value. In turn, this leads to new productions of nature which become dominated by second nature in all three of the senses implied by Smith. So, for instance, Smith writes 'capital stalks the earth in search of material resources', transforming nature into an 'appendage of the production process' (Smith, 1984: 49) and leaving 'no original relation with nature unaltered, no living thing unaffected' (Smith, 1984: xiv). Moreover, the historically emergent and highly

abstract exchange-value relations of capitalism become naturalized as 'second-nature' and come to produce not only non-human nature but also human nature. Capitalist production is, Smith argues, now 'squarely at the centre of nature' (Smith, 1984: 65).

From Chapter 3 of the book, Smith turns his attention to the issues of space and the concept of uneven development. He starts by considering the *'production of space'*, a notion which he notes is 'logical corollary of the production of nature' (Smith, 1984: 66), although he adds that in this second part of the book he is keen to develop this concept independently from his discussion of the production of nature, before seeking to drawing links between the two productions. This having been said, it is clear that Smith adopts very similar lines of argument, outlining what he sees are key problems in existing theories before drawing on Marxist concepts to develop an understanding which, in his view, rectifies many of these problems.

The initial problem identified by Smith is the presence of 'deeply engrained and commonly held prejudices concerning space', whereby space is seen as 'a field of activity or as a container' and where there is 'a rather mechanical integration of space and society' (Smith, 1984: ix). This understanding, Smith argues, embodies elements of two concepts of space known formally as *absolute* and *relative* space, elements which, as with the scientific and poetic concepts of nature, exist in a tangled inter-relationship. Furthermore, as with nature, Smith argues that these concepts emerged in association with changes in the modes of human productive activity related to the rise of market exchange and capitalism. He suggests that the concept of absolute, or *Newtonian space*, whereby space is seen as 'a universal receptacle in which objects and events occur' and which thereby acts as 'a frame of reference, a co-ordinate system … within which all reality exists' (Smith, 1984: 68), appeared in association with the practices such as cadastral surveying and the rise of commodity exchange which fostered abstraction from both use and from the immediacy of place. Similarly, mathematical concepts of relativistic space associated with, for instance, Einstein's theory of relativity, are seen by Smith to be expressions that relate to the development of capitalist production; and also the associated development of 'second nature' whereby society came to be viewed as separate from nature and notions of distinct natural and social spaces emerge abstracted from each other and from locations in physical, Newtonian space.

Smith goes on to argue that not only have these concepts of space emerged out of production for exchange and capitalist production, but they have been integral to the workings of these modes of production. Focusing primarily on capitalist production and drawing heavily on ideas associated with Marx and his 'labour theory of value', Smith suggests that capitalist production involves the production of both absolute and relative space. Absolute space, he argues, is significant in that concrete labour

processes require a particularity of spatial attributes: particular people, raw materials, and technologies combine in specific locations to produce commodities. However, within capitalism this particularity is always situated in – and produced out of – wider relations: for instance, the product being produced will be offered up for sale to a diversity of purchasers in a wide range of locations, and the people who produce it will be 'hired' in competition with other people. Smith further notes tensions between the capitalist production of these spaces, suggesting that while there is a general historical movement towards relative space, epitomized in the globalization of capitalist production and the increasing mobility of capital, absolute space is not eliminated but instead moves from spaces of 'first nature' to spaces of 'second nature'. In other words, whilst capitalism increasingly relativizes nature through valuation via exchange values, it also creates new particularities of space by, for instance, striating space through social relations such as property and state territories, and through the fixity required by some forms of investment of capital (a point which is quite central to Smith's 'rent-gap' theory of gentrification – see Smith, 1996). Within capitalist production there are, Smith argues, inherently contradictory tensions between geographical differentiation and equalization, tensions whose resolution, he suggests, provide the basis for understanding the basis of uneven development.

For Smith there is, however, one further issue which needs to be addressed in order to 'completely specify' (Smith, 1984: 131) the processes underlying uneven development, and that is the issue of scale. Once again the focus for Smith is production; hence *Uneven Development* involves not only consideration of the production of nature and the production of space, but also requires consideration of the *'production of scale'*, with this last construction itself very much being produced out of the first two creations. Drawing on his discussion of the production of space, for instance, Smith

Table 9.2 The production of scale within capitalist production

Spatial scale	Dominant spatial tendency	Key features
Global	Equalization	Universalization of the wage–labour relation through both political and economic integration encourages equalization. However, differentiation does occur, principally through the differential value of labour power which results in an international division of labour.
Nation	–	Not directly conditioned by capital differentiation and equalization, but rather is the product of series of political deals, compromises and conflicts. Key impetus comes from the competition between different capitalist organizations, with the nation–state often acting to protect the collective interests of capital within its boundaries, plus acting to defend territories militarily and to regulate and guarantee the maintenance of the working class.
Urban	Differentiation	Manifestation of the centralization of capital which creates urban space as an 'absolute space of production'. Extent determined by extent of local labour market and limit of daily commute within which equalization works. However, differentiation predominates through ground–rent system which sorts competing land uses into different areas.

Source: Based on Smith (1984)

suggests that the dialectic of differentiation and equalization of capitalist production is bound up with concepts of absolute and relative space and first and second natures, producing three 'primary' scales of spatiality: the urban, the nation-state and the global. Importantly, while these terms have been routinely employed within geographical analysis, Smith notes the arguments of Taylor (1982) concerning the need to view them not as pregiven, natural entities, but to reflect on their role in the social production of space. For Smith the existence of these scales stems principally from the role they play in relation to the processes of differentiation and equalization. In particular, he suggests that the urban and the global scales respectively 'represent the consummate geographical expression of the contradictory tendencies toward differentiation and equalization' (Smith, 1984: 142), while nation-states are viewed as 'differentiated absolute spaces' created as a result of 'historical deals, compromises and wars' (see Table 9.2).

Neil Smith ends *Uneven Development* by highlighting how his abstract analysis might help illuminate aspects of the uneven development of the world in the 1970s and early 1980s, plus he discusses some of the omissions of his analysis and its inherently political dimension. For Smith this period witnessed a restructuring of geographical space which was 'more dramatic than any before', and involved changes such as '[d]eindustrialization and regional decline, gentrification and extrametropolitan growth, the industrialization of the Third World and a new international division of labour, intensified nationalism and a new geopolitics of war'. The range of issues, the explicit call for recognition of the political dimensions of academic study and the extensive utilization of ideas derived from the work of Marx can all be seen to situate *Uneven Development* and its author in the emergence of radical/Marxist geography: indeed Peet (1998: 93) cites the book as 'symptomatic of the best of Marxist geography' while Castree (2000: 269) argues that it 'was a major contribution to the development and consolidation of Marxist geography as a leading postpositivist paradigm'.

Smith wrote *Uneven Development* soon after completing a doctorate under the supervision of the renowned Marxist geographer David Harvey at Johns Hopkins University, Baltimore, having previously studied at the University of St Andrews, Scotland. Corbridge (1987) argues that *Uneven Development* exhibits a clear debt to Harvey, particularly his *Limits to Capital* (see Castree, Chapter 8 this volume). Smith himself, as noted earlier, explicitly argued that his book drew upon ideas developed in his doctorate which focused on one aspect of the restructuring of geographical space mentioned in *Uneven Development*, namely gentrification, whereby central areas of cities were seen to be being redeveloped for and in some cases by middle class residents. By the time of publication of *Uneven Development,* Smith had already published several articles which sought to demonstrate the link between gentrification and the movement of capital associated with uneven development (e.g. Smith, 1982) and subsequently Smith has developed an extensive corpus of work on gentrification, much of which continued to stress how gentrification could be seen as the product of spatially more general processes associated with uneven development (see especially Smith, 1996). Prior to *Uneven Development* Smith had also written on Marxist's conceptualization of nature (Smith and O'Keefe, 1980) and on the theory and practices employed in geography (Smith (1979). Subsequent to its publication Smith has continued to explore both strands of work (e.g. Smith, 1987, 1992b) as well as continuing arguments relating to the production of scale (Smith, 1992a).

Reception and evaluation

Uneven Development emerged at a time of considerable debate about the purpose and form of geography, and given this context it is not unsurprising that much of the commentary on the book has focused on the contribution that it made to the establishment of a Marxist perspective and the degree to which it exhibited some of the perceived strengths and weaknesses of this approach (e.g. Corbridge, 1987; Sack, 1987; D. Smith, 1986; Castree, 2000; Swyngedouw, 2000). In retrospect, Smith's book might be seen to enact what now is often described as a 'modernist' approach or attitude, whereby theory is seen to revolve around the production of concepts which 'correspond to central phenomena or processes around which the world is seen to operate' (Phillips, 2005: 56). *Uneven Development* visibly aspired to be 'grand theory', and as such can be subjected to criticism for being both totalizing and insensitive to difference, although, as Castree (1995: 30) notes, the Marxism of Smith does temper its explanatory-diagnostic impulse by recognizing that knowledge is itself produced within historically, and spatially specific, contexts, and hence 'embodies a moment of self-reflectivity concerning its claims to truth and reason'. For Castree, however, this self-reflective impulse is not taken far enough and hence the work of Smith still enacts an overly modernist attitude.

Smith has, in part, recognized some deficiencies in his analysis, recognizing that gender relations may have dynamics which cannot be read directly off the dynamics of production as implied in *Uneven Development*. He has also subsequently bequeathed greater agency to social identity and subjectivity (Smith, 1996a; see also see Marston, 2000). This is not to say, however, that *Uneven Development* has been consigned to gather dust on people's shelves, to be pulled out only as an illustration of a now unfashionable approach. Quite the opposite, as the 1990s saw the book re-published (Smith, 1990) and become widely cited, albeit arguably more for some of the terms and concepts it invoked rather than for its over-arching, and complexly inter-related, theories of uneven development. In particular, Smith's concepts of the production of nature and scale have become the focus of considerable research and debate (see for instance Castree, 1995, 2002; Marston, 2000). Once again, not all of the ensuing commentary has fully endorsed the arguments of *Uneven Development*. Castree (1995: 20), for instance, has argued that whilst the book represented a key intervention in the debate about capitalism and nature, Smith's concept of the production of nature seems to imply that capitalism 'determines every aspect of nature', leaving no room for any 'agency of nature', whereby entities and processes might be conditioned by something other than human labour, either in general or by the particularities of capitalist production. To return to the language of Smith, perhaps first natures have not been totally erased by capitalist second nature. Similarly, with respect to the production of scale, questions have been raised both about the degree to which capitalist production is the sole producer of scale, and indeed whether scale itself exists (see Marston, 2005; Smith, 2003).

More than 20 years after its publication, *Uneven Development* is still clearly the subject of interest and debate. Indeed Castree (2000: 269) has suggested that the 'power and relevance' of the book appear to have increased rather than diminished in the years since its initial publication. Developments such as genetic engineering and cloning, for instance, can be viewed as indicators of the production of humanly produced 'second natures', whilst globalization can be viewed as a movement into relativistic space which has not only continued but accelerated since the 1980s. Furthermore, as Smith (2000) notes, globalization raises the issue of scale quite directly, not least in relation to the power and significance of the nation-state. *Uneven Development* is hence quite clearly a book of great contemporary relevance, although, as my

own initial encounters with the text testify, it is not necessarily a very easy book to get to grips with. Not only does it challenge many established perceptions of how the world is structured – for instance, in terms of distinct realms of society and nature – but, as I have tried to illustrate here, uses terms in multiple and relational ways. So for instance, Smith not only differentiates first and second natures in conjunction with his abstract/concrete distinctions (Figure 9.1), but in turn links these to both use- and exchange-value and absolute and relative space distinctions. Hence, in market exchange societies production is seen as increasingly orientated towards production for exchange and placed in the abstract framework of absolute space, although retaining a grounding in the concrete materiality of first nature. In capitalist production societies, nature itself became increasingly a second nature product in relation to exchange valuation, whilst production in general is increasingly located in relativistic space.

By contrast to such 'total theory' where 'everything hangs together with everything else' (Castree, 2000: 267), subsequent researchers have tended to select particular components of the book for discussion and research, most notably the concepts of 'production of nature' and 'production of scale', without really endorsing, or arguably perhaps even being aware of, the book's overall perspective. The production of nature, space and scale have come to be phrases of great significance in contemporary human geography, although studies associated with them have been informed by a number of perspectives, such as actor-network theory and social/cultural constructionism, that might be seen as quite antithetical to the Marxist theorization advanced by Smith. There have been some claims that there may be scope for some rapprochement between these approaches (see Castree, 2002), as well as some calls to reflect of what might have been lost within recent appropriations of *Uneven Development*. Swyngedouw (2000: 268), for instance, suggests that the book is distinctive and 'even more powerful today than at the time of its writing' in its willingness to 'grapple with "big issues"'. Castree (1995: 27) similarly suggests that there are still important contributions to be made by a Marxism which, whilst taking seriously 'limited discursive worlds' and local environmental struggles and problems, still sees value in building 'a larger project which situates those problems and joins their antagonism into a more global view of and struggle over nature's (and society's) creative destruction'.

Secondary sources and references

Castree, N. (1995) 'The nature of produced nature: materiality and knowledge construction in Marxism', *Antipode* 27: 12–48.

Castree, N. (2000) 'Classics in human geography revisited', *Progress in Human Geography* 24: 268–271.

Castree, N. (2002) 'False antethesis? Marxism, nature and actor-networks', *Antipode* 34: 111–146.

Castree, N. and Braun, B. (2001) *Social Nature: Theory, Practice and Politics*. Oxford: Basil Blackwell.

Corbridge, S. (1987) 'Review: Neil Smith, Uneven development: nature, capital and the production of space', *Antipode* 19: 85–87.

Harvey, D. (1982) *The Limits to Capital*. Oxford: Basil Blackwell.

Marston, S. (2000) 'The social construction of scale', *Progress in Human Geography* 24: 219–242.

Marston, S., Jones, J.P. III and Woodward, K. (2005) 'Human geography without scale', *Transactions, Institute of British Geographers* 30: 416–432.

Marx, K. (1976) *Capital: A Critique of Political Economy, volume 1*. Harmondsworth: Penguin.

Peet, R. (1998) *Modern Geographical Thought*. Oxford: Blackwell.

Phillips, M. (2005) 'Philosophical arguments in human geography', in M. Phillips (ed.) *Contested Worlds: An Introduction to Human Geography*. London: Ashgate.

Phillips, M. and Mighall, T. (2000) *Society and Exploitation Through Nature*. Harlow: Prentice Hall.

Sack, D. (1987) 'Review: Uneven development: nature, capital and the production of space by Neil Smith', *Geographical Review* 77: 130–132.

Sayer, A. (1984) *Method in Social Science: A Realist Approach*. London: Hutchinson.

Smith, D. (1986) 'Review: Uneven development: nature, capital and the production of space by Neil Smith', *Transactions of the Institute of British Geographers*, new series 11: 253–254.

Smith, N. (1979) 'Geography, science and post-positivist modes of explanation', *Progress in Human Geography* 3: 356–383.

Smith, N. (1982) 'Gentrification and uneven development', *Economic Geography* 58: 139–155.

Smith, N. (1984) *Uneven Development: Nature, Capital and the Production of Space*. Oxford: Basil Blackwell.

Smith, N. (1987) 'Academic war over the field of geography': the elimination of geography at Harvard, 1947–1951, *Annals of the Association of American Geographers* 77: 155–172.

Smith, N. (1990) *Uneven Development: Nature, Capital and the Production of Space* (2nd edition). Oxford: Basil Blackwell.

Smith, N. (1992a) 'Contours of a spatialized politics: homeless vehicles and the production of geographical scale', *Social Text* 33: 55–81.

Smith, N. (1992b) 'History and philosophy of geography: real wars, theory wars'. *Progress in Human Geography* 16: 257–271.

Smith, N. (1996) *The New Urban Frontier: Gentrification and the Revanchist City*. London: Routledge.

Smith, N. (1996a) 'Spaces of vulnerability: the space of flows and the politics of scale', *Critique of Anthropology* 16: 63–77.

Smith, N. (2000) 'Socializing culture, redicalizing the social?', *Social and Cultural Geography*, 25–28.

Smith, N. and O'Keefe, P. (1980) 'Geography, Marx and the concept of nature', *Antipode* 12: 30–39.

Smith, R.G. (2003) 'World city topologies', *Progress in Human Geography* 27: 561–582.

Swyngedouw, E. (2000) 'Classics in human geography revisited', *Progress in Human Geography* 24: 266–268.

Taylor, P.J. (1982) 'A materialist framework for political geography', *Transactions, Institute of British Geographers* 7: 15–34.

10 SPATIAL DIVISIONS OF LABOUR (1984): DOREEN MASSEY

Nick Phelps

Space can be … conceptualised as the product of the stretched-out, intersecting and articulating social relations of the economy. (Massey, 1984: 2)

Introduction

At the outset of this chapter we can say that, like all books (academic or literary), *Spatial Divisions of Labour* is one 'which requires evaluation at different levels' (Smith, 1986: 189). It is curious to think that it is at one and the same time deeply immersed in the specific concerns of industrial location – a term now rarely used in mainstream economic geography – and yet punctuated by profound observations on the philosophy of human geography. It is at once a critique of existing location theory and economic geographical work and yet also a new approach to understanding uneven economic development. It is a theoretical *tour de force* and yet also a detailed empirical exploration of that theoretical approach. Among a group of like-minded scholars its importance was immediate and prompted a major academic research programme – the Changing Urban and Regional Systems (CURS) initiative – funded by the UK government's Social Science Research Council (now the Economic and Social Research Council). Yet it has also had a profound longer-term impact in shaping new academic research, presenting new points of departure for work on geography

and gender, geometries of power as well as prefiguring interest in, for example, labour movement strategies and the institutional and cultural basis of capitalism. It is these multiple impacts that have ensured *Spatial Divisions of Labour*'s stature and its enduring appeal. Moreover, in all of this, something of the author's character is revealed right from the first until the very last page.

After having read geography at Oxford University (1963–1966), Doreen Massey went on to study for an MA at University of Pennsylvania, Philadelphia (1971–1972) as a conscious decision to immerse herself in positivist social science in general, and neoclassical economics, location theory and regional science in particular. In the years following, while working at the Centre for Environmental Studies, Massey was to develop the various strands of a critique of established positivist and behavioural location theory (Massey, 1977, 1978). This was also a time of profound restructuring of the UK economy – a period sometimes referred to as de-industrialization. As Massey herself noted, the book was written at a moment that allowed reflection on major transformations occurring since the 1960s. It is the 'deep' understanding of these processes of restructuring in the UK, their industrial specificity (Massey and Meegan, 1982) and diverse impacts on cities and regions (Massey and Meegan, 1979, 1978) that forms the empirical substance of *Spatial Divisions of Labour*. In this regard, and 'as an analysis of employment changes in post-war Britain', it is 'full of fascinating insights' (Smith, 1986: 189). However, the book today may appear slightly curious with its many references to a branch of human geography – industrial geography – now largely disregarded. In this and in other respects, *Spatial Divisions of Labour* is, as we have come to realize through the work of Trevor Barnes (2004), most definitely a product of a particular time and place. And yet, through its clarity, force of argument and novelty, *Spatial Divisions of Labour* has gained sufficient longevity to be used and cited in markedly different circumstances (sometimes in a way that Doreen Massey herself might have guarded against). As such, while Doreen Massey had already established herself as a leading and highly innovative economic geographer when *Spatial Divisions of Labour* was published in 1984, its publication (and subsequent reissue in 1995) cemented her reputation in the discipline as a whole.

The book and its arguments

From the start, Massey's *Spatial Divisions of Labour* was much more than an industrial geography of the UK. As she argued, 'the study of industry and production is *not* just a matter of "the economic" but rather that economic relations and phenomena are themselves constituted on a wider field of social, political and ideological relations' (Massey, 1984: 7, original emphasis). As such, *Spatial Divisions of Labour* prefigured recent institutional and cultural 'turns' in economic geography. And, as David Smith noted at the time, 'the book goes well beyond industrial and employment matters [to offer] a major methodological redefinition of human geography' (Smith, 1986: 190). It both set out to develop a new theory of uneven geographical development at the same time as it exemplified those concerns empirically. What we have is 'a densely argued book that constantly swings from abstract theory to finely-honed empirical examples, culminating in a philosophical claim about the very nature of social life' (Clark, 1985: 291). If at some points *Spatial Divisions of Labour* is grounded in the specific concerns of industrial geography, at others it is intervening in the most abstract concerns of the philosophy of human geography. At the heart of *Spatial Divisions of Labour* is a desire, as Massey later clarified, to reconceptualize space as 'the product of the stretched-out, intersecting and articulating social relations of the economy' (Massey, 1995: 2).

As I have already alluded to briefly, *Spatial Divisions of Labour* was steeped in a thorough understanding of the turmoil of the UK economy during the 1960s and 1970s. Somewhat unusually for a book that offers a major re-conceptualization of a sub-disciplinary field, it is also firmly grounded in a rich empirical knowledge of its subject. In fact, as Doreen Massey has since recounted, the inclusion of largely empirical chapters – albeit that they are central to her project of critique and reconstruction and to her dialectical movement between theory and evidence – was seen at the time to have been something of a potential weakness (for example, it was suggested to the author at the time that the empirical examples in the book might come to be regarded as ideal types and hinder the wider impact of the book). These largely empirical elements focused on, for example, the evolving spatial structures associated with the electronics, clothing and footwear and service industries in the UK, upon the class and gender relations in coalfield communities (of south Wales, the northeast of England, central Scotland, and Cornwall), and upon broader patterns of uneven development, class formation and politics in the UK as a whole. Nevertheless it was the *theoretical* content of *Spatial Divisions of Labour* and other works by radical geographers that at the time elevated geography among related disciplines such as sociology, planning and even economics – with the majority of those involved in the CURS initiative, for example, being drawn from these disciplines outside of geography (Cooke, 2008).

In the manner of its creation, *Spatial Divisions of Labour* also poses some interesting questions of much contemporary academic work in human geography. Recent debates in academic journals have lamented the detachment of human geographical research from and its lack of impact upon policy and politics in the UK in particular (see, for example, Dorling and Shaw, 2002; Martin, 2001, 2002; Massey, 2001, 2002). Yet even at the time and certainly by today's standards, *Spatial Divisions of Labour* is very unusual in being forged from the author's life-long engagement with policy and politics – a theme that was to form the title of a subsequent volume co-edited by Massey (Massey and Meegan, 1985).

The Academy – as a central institution of civilizations – has several important functions. At times it is the centre of the pursuit of abstract knowledge far removed from everyday concerns – the subject of Herman Hesse's novel *The Glass Bead Game*. At other times it has been the centre of societal critique and resistance. Rarely, it seems, is academia involved in both critique and reconstruction. Massey's *Spatial Divisions of Labour* is one such exception and it owes so much to the author's close involvement with practical political and policy concerns. These concerns continue today in the guise of, for example, her involvement with the think tank 'Catalyst' and questions of regional policy (Amin *et al.*, 2003). Prior to the writing of *Spatial Divisions of Labour*, Massey had been closely involved with the Greater London Council's *London Manufacturing Strategy* developed during the late 1970s and early 1980s. Indeed, there is a case for arguing that the clarity, power and persuasiveness of this most inscrutable of academic books owes much to

the rigour required explaining concepts in arenas and for audiences outside academia. Hence the book speaks to the process not only of understanding industrial restructuring on the ground in London and elsewhere but also to the immediate and pressing need to find practical policy solutions to the resultant social ills. While *Spatial Divisions of Labour* and the locality studies that ensued under the CURS initiative came at a propitious time in relation to theoretical and methodological developments in geography and elsewhere in the social sciences, it was developments in the wider world that they were most in tune with (Massey, 1995).

Spatial Divisions of Labour sought to conceptualize the geography of employment in terms of the social relations of production. To this end it advanced an approach centred on three more concrete conceptual innovations: 'place in economic structure' – the relation of a particular industry or economic sector to other (in Marxist parlance) departments; 'the organisational structure of capital' – a term used to capture dimensions of ownership and capital concentration apparent in a particular sector of the economy; and the various 'spatial structures' that may characterize different industries at different times. It is the latter conceptual category that *Spatial Divisions of Labour* made most famous when identifying the 'locationally-concentrated', 'cloning' and 'part-process' spatial structures that tended to be associated with particular industries at particular times and in particular settings. These spatial structures were closely related to – though not determined by – the exigencies of different labour processes. They were also considered by Massey as merely examples that did not exhaust the range of existing or potential spatial structures. In this respect then, we see Massey's interest in and determination to 'formulate an approach to conceptualising … the arguably endless adaptability and flexibility of capital' (Massey,

1984: 69). However, such was the novelty and power of these categories that they have been and, to some extent, continue to embody rather static reference points or archetypes for researchers.

Reception and impacts

It is easy to forget that a book of the stature of *Spatial Division of Labour* nevertheless also attracted no small measure of detailed evaluation and critique at the time. Despite general approval from most reviewers, a slew of more or less important criticisms were levelled at the book. Massey's bold attempt to bridge the nomothetic–idiographic epistemological divide in human geography was itself the single largest cause of criticism. Massey herself reflected years later in the second edition how the book was criticized both as Marxist and yet not Marxist enough (Massey, 1991: 297). So for John Lovering, for example, Massey's imposition of Wright's abstractions on class onto concrete situations produced an economistic feel to the analysis (Lovering, 1986). However, the weight of criticism at the time fell squarely on the side of the book being too concerned with the uniqueness of place and hence not being principled enough in Marxist or realist terms. Smith (1986: 180) argued that 'at the heart of the analysis is the uniqueness of place', with Massey viewing 'causal relations' as enabling rather than determining (see Massey, 1995: 4). In the opening chapters Massey was concerned to avoid an essentialist Marxist account of uneven development, developing her criticisms of ideas that had surfaced and were popular at the time, arguing that 'it is not enough to point to the stage reached in the development of capitalist relations in order to understand the complexity of spatial structures' (Massey, 1984: 81). Although not an explicit target of Massey's critique in this respect, Nigel Thrift was able to observe how

'one of the oddest omissions in the book is any mention of the work of David Harvey' (Thrift, 1986: 148) – odd since David Harvey was and remains *the* major Marxist figure in human geography (see Woodward and Jones, Chapter 15 this volume). However, Harvey was not to stay silent on the matter for long, lamenting the extent to which *Spatial Divisions of Labour* was 'laden down with the rhetoric of contingency, place, and specificity' to the extent that Marxian categories became inert (Harvey, 1987: 373). This line of argument was pursued in various guises by several other commentators, notably those working in the realist tradition that had also become popular at the time. For Gordon Clark (1985: 292) what seemed lacking was a 'macro sense of the structures of the regional system from an economic perspective'. Warde (1985) had suggested that *Spatial Divisions of Labour* was 'highly indeterminate' in its analysis, while Cochrane identified how 'there do not seem to be any general rules – or necessary relations – at all' such that 'we are left with a process of infinite regress, in which the model is presented, only to recede into the distance every time we try to use it' (Cochrane, 1987: 361). This was not the intention of the book although it was something that Massey subsequently acknowledged as a potential danger in approaches that attempt to deal seriously with industrial and locational specificity (Massey, 1995: 320).

Perhaps the most immediate impact of the book was to set in train a major UK research programme – the CURS initiative (Cooke, 1989; Harloe *et al.*, 1990) – the machinations of which also condensed many of the above concerns. As Lovering notes 'in the three hundred pages devoted to the "specificity of place" there is no concise or complete conceptualization of the locality' (Lovering, 1986: 71). The search for such a conceptualization occupied much of the research effort at the time both within and without the various locality studies. As we have already noted,

Massey studied during a time of transition between the long-dominant regional geography tradition and the rapid succession of nomothetic approaches that characterize the 1960s and 1970s. These twin influences are apparent in *Spatial Divisions of Labour*. Massey the person, and *Spatial Divisions of Labour* the book, were to become icons within Marxist human geography and yet David Smith was able to articulate a sense of the familiar noting how 'there is more than a hint of a revitalized regional geography' (Smith, 1986: 190). The theme was taken up by Lovering who suggested that 'the case studies also show how information produced by empirical studies of a quite traditional kind may be given a new meaning when situated in a more coherent theoretical framework' (Lovering, 1986: 70). For others the attempt to navigate the conceptual straits between nomethetical and idiographic approaches was less satisfactory since the locality studies that *Spatial Divisions of Labour* spawned 'will do little more than repeat the empiricist locality studies of an earlier generation which examined individual places for their own sake' (Smith, 1988: 62).

Both *Spatial Divisions of Labour* and the locality studies commissioned in its wake appeared at a time of rapid development of theoretical and philosophical approaches in human geography that included not only radical/Marxist approaches that had surfaced by the late 1970s but also in particular the realist and structurationist approaches that had become popular by the 1980s. Thus it was that most of the sympathetic critique and reconceptualization of locality studies came from a realist perspective, culminating in Duncan's (1988) 'three levels of locality' effects and in Sayer's (1989) defence of the 'new regional geography' as distinct from the older regional geography tradition.

The tension generated by the dominant nomothetic emphasis on structural processes on the one hand and the idiographic emphasis on the uniqueness of place on the other hand is one that begs important questions of the historical transformation of places. Massey herself likened the process to the layering of rounds of investment whereby 'spatial structures of different kinds can be viewed historically ... as evolving in a succession in which each is superimposed upon, and combined with, the effects of the spatial structures which came before' (Massey, 1984: 118). The localities research that followed was to occupy much of its research effort with such 'rules of transformation'. In doing so, one inappropriate metaphor achieved a prominence and illustrates clearly the potential for unintended consequences to flow even from the clearest of academic expositions. The notion of layers of investment is now intimately and erroneously associated with a 'geological metaphor' originally drawn by Warde (1985). Massey herself felt it necessary to point out 'geology is *not* an appropriate way of envisaging the layering process I had in mind ... surely the notion of the *combination* of layers is very ungeological' (Massey, 1995: 231, original emphasis). An alternative, more accurate metaphor of the dealing of hands of cards was suggested by Gregory (1989).

Critiques and reappraisals

Inevitably, a book as ambitious and wide-ranging as *Spatial Divisions of Labour* was likely to neglect quite specific aspects of the phenomena it was concerned with. Several detailed limitations of the book were highlighted by reviewers at the time. Thrift (1986) and Warde (1985), for instance, highlighted the limited attention paid to the role of the reproduction of labour in the book. This is something that Massey conceded in the second edition, noting the book 'never meant to encompass uneven development in every sphere of life' (Massey, 1995: 334). *Spatial Divisions of Labour* focused on the spatial structures produced from the technical divisions of labour internal to companies. In this respect Massey was careful to note that the 'locationally-concentrated', 'cloning' and 'part-process' spatial structures concentrated on in the book were not in any sense exhaustive of possibilities. Yet *Spatial Divisions of Labour* clearly had an emphasis on the latter two spatial structures and partly as a function of this tended to underplay the inter-firm linkages or social division of labour often associated with the single plant spatial structure. This was to become a major focus of academic enquiry during the late 1980s and early 1990s as academics rediscovered the phenomena of industrial districts and agglomeration. One of the major figures involved in this enterprise – Michael Storper – was to raise this criticism (Storper, 1986) (see also Coe, Chapter 17 this volume). It is a limitation that Doreen Massey readily conceded, noting how 'extending the … analysis … to all the various levels of economic relations, especially including those between firms, would give a much fuller, and more complex and … possibly contradictory, picture of spatial divisions of labour' (Massey, 1995: 339).

Spatial Divisions of Labour also arguably belongs to an increasingly rare corpus of academic geographical writing in which ideas are expressed with clarity. As was noted at the time (Smith, 1986; Thrift, 1986), *Spatial Divisions of Labour* is a very well-written book that is free from jargon. The clarity of its expression notwithstanding, it is also true that the book is at times an uncomfortable read. It is uncomfortable for a conceptual rigour that exposes the laziness in preceding academic research in the fields of locational analysis, regional economic development, and even some strands of Marxist economic geography. What comes across repeatedly is an insistence and a determination in the presentation and working through of key issues that avoid easy or fallacious solutions.

As one might expect, such a book and its style are the product of a lively and determined mind. So, David Smith, for example, was able to observe how 'Some … will be intrigued by the imagery of class conflict ("battle", "attack", "assault", "weapons") – as if Massey was demonstrating that a confrontational stance, like the struggle of labour itself, is not a male preserve' (Smith, 1986: 190). Ought we to be surprised that the steely qualities needed for Doreen Massey to assert herself as an academic leader of immense stature within what was and, to an extent, remains a male-dominated profession should appear at moments on the page and in person? These are qualities that every academic would surely recognize in their engagement with the sometimes alienating machinery of academic life, although they are, along with altogether less admirable traits, ones rarely explicitly acknowledged in the academic community.

It is also salutary to remember that *Spatial Divisions of Labour*'s impact at the time was, unlike geography's quantitative revolution (Barnes, 2004), not borne of a deliberate or concerted effort on the part of the author to bend the academic machinery in its service – the sort of effort that has now become perhaps too readily embraced in the practice of geography. Whilst Doreen Massey assumed a very influential position as a result of its publication, she was unable to claim any special efforts on her part to garner impact for *Spatial Divisions of Labour*. If anything, her concerted efforts to promote progressive geographical thought has come outside academia in the realm of policy and politics, since Doreen Massey has been a regular contributor to periodicals such as *Marxism Today* and *Soundings*. Instead, the success of *Spatial Divisions of Labour* and the CURS initiative owe much to the sheer variety and geographical scope of the associations that Doreen Massey had struck up across several overlapping academic and policy groupings as well as the sympathy of established geographical figures such as Michael Wise to the aspirations of radical geographers.

The power of Doreen Massey's *Spatial Divisions of Labour* rests on its desire to address *the* fundamental schism in human geography. It navigates a course between an idiographic tradition – the description of place – on the one side and the nomothetic outlook – of the study of process – on the other. Its beauty and power lies in the author's calm, clarity of purpose and expression and tenacity in navigating these waters. It is the middle ground – this tension between two rather distinct approaches in human geography – that arguably informs much of the most interesting work within human geography which followed.

Secondary sources and references

Amin, A., Massey D. and Thrift, N. (2003) 'Decentering the nation: A radical approach to regional inequality', *Catalyst Paper 8*. London: Catalyst.

Barnes, T. (2004) 'Placing ideas: genius loci, heterotopia and geography's quantitative revolution', *Progress in Human Geography* 28: 565–595.

Clark, G.L. (1985) 'Review of Spatial Divisions of Labour', *Economic Geography* 61: 290–292.

Cochrane, A. (1987) 'What a difference the place makes: the new structuralism of locality', *Antipode* 19: 354–363.

Cooke, P. (ed.) (1989) *Localities: The Changing Face of Urban Britain*. London: Unwin Hyman.

Cooke, P. (2008) 'Locality debates', in R. Kitchin and N. Thrift (eds), *The International Encyclopedia of Human Geography*. Oxford: Elsevier.

Dorling, D. and Shaw, M. (2002) 'Geographies of the agenda: public policy, the discipline and its (re)'turns', *Progress in Human Geography* 26: 629–646.

Duncan, S.S. (1988) 'What is locality?', in R. Peet and N. Thrift (eds), *New Models in Geography*. London: Allen & Unwin.

Gregory, D. (1989) 'Areal differentiation and post-modern human geography', in D. Gregory and R. Walford (eds), *Horizons in Human Geography*. London: Macmillan.

Harloe, M., Pickvance, C. and Urry, J. (1990) (eds). *Place, Policy and Politics: Do Localities Matter?* London: Unwin Hyman.

Harvey, D. (1987) 'Three myths in search of a reality in urban studies', *Environment & Planning D, Society & Space* 5: 367–376.

Lovering, J. (1986) 'Book review: Spatial Divisions of Labour', *Urban Studies* 23: 70–71.

Martin, R. (2001) 'Geography and public policy: the case of the missing agenda', *Progress in Human Geography* 25: 189–209.

Martin, R. (2002) 'A geography for policy or a policy for geography? A response to Dorling and Shaw', *Progress in Human Geography* 26: 642–644.

Massey, D.B. (1977) 'Towards a critique of industrial location theory', in R. Peet (ed.), *Radical Geography*. London: Methuen.

Massey, D.B. (1978) 'A critical evaluation of industrial location theory', in F.E.I. Hamilton and J.C.R. Linge (eds), *Spatial Analysis, Industry and the Industrial Environment: Industrial Systems*. Chichester: Wiley.

Massey, D.B. (1979) 'In what sense a regional problem?', *Regional Studies* 13: 233–243.

Massey, D.B. (1984) *Spatial Divisions of Labour: Social Structures and the Geography of Production.* Basingstoke: Macmillan.

Massey, D.B. (1991) 'The political place of locality studies', *Environment & Planning A* 23: 267–281, reprinted in Massey, D.B. (1994) *Space, Place and Gender.* Cambridge: Polity Press.

Massey, D.B. (1995) *Spatial Divisions of Labour: Social Structures and the Geography of Production* (2nd edition). Basingstoke: Macmillan.

Massey, D.B. (2001) 'Geography on the agenda', *Progress in Human Geography* 25: 5–17.

Massey, D.B. (2002) 'Geography, policy and politics: a response to Dorling, and Shaw', *Progress in Human Geography* 26: 645–646.

Massey, D.B. and Meegan, R. (1978) 'Industrial restructuring versus the cities', *Urban Studies* 15: 273–289.

Massey, D. and Meegan, R. (1979) 'The geography of industrial reorganisation: the spatial effects of the restructuring of the electrical engineering sector under the Industrial Reorganisation Corporation', *Progress in Planning* 10 (3): 155–237.

Massey, D.B. and Meegan, R. (1982) *The Anatomy of Job Loss: The How Why and Where of Employment Decline.* London: Methuen.

Massey, D.B. and Meegan, R. (1985) *Politics and Method: Contrasting Studies in Industrial Geography.* London: Methuen.

Sayer, A. (1989) 'The 'new' regional geography and problems of narrative', *Environment & Planning D, Society and Space* 7: 253–276.

Smith, D.M. (1986) Review of Spatial Divisions of Labour, *Regional Studies* 20: 189–190.

Smith, N. (1987) 'Dangers of the empirical turn: some comments on the CURS initiative', *Antipode* 19: 59–68.

Storper, M. (1986) 'Review of Spatial Divisions of Labour', *Progress in Human Geography* 455–457.

Thrift, N. (1986) 'Review of Spatial Divisions of Labour', *International Journal of Urban and Regional Research* 10: 147–149.

Warde, A. (1985) 'Spatial change, politics and the division of labour' in D. Gregory and J. Urry (eds), *Social Relations and Spatial Structures.* Basingstoke: Macmillan, pp. 190–212.

11 GEOGRAPHY AND GENDER (1984): WOMEN AND GEOGRAPHY STUDY GROUP

Susan Hanson

What we argue for ... is not ... an increase in the number of studies of women *per se* in geography, but an entirely different approach to geography as a whole. Consequently, we consider that the implications of *gender* in the study of geography are at least as important as the implications of any other social or economic factor which transforms society and space. (WGSG, 1984: 21)

Introduction

Imagine a world outside your classroom in which women's subordinate status is vividly evident through gender differences in educational attainment (years of formal education), employment (job type, hours of employment, wages), and political power (presence/absence in governmental decision-making positions). Imagine a world inside academia in which geography journals and university geography curricula are practically devoid of the words 'gender' and 'women' and in which all instructional materials implicitly, if not explicitly, assume that all salient geographical actors are male. Imagine university geography teaching staffs, graduate student cohorts, and national geography conferences composed almost entirely of white men. Such was the world of the early 1980s in which nine members[1] of the Women and Geography Study Group collectively conceived and drafted *Geography*

and Gender: An Introduction to Feminist Geography (1984).

The text and its authors

Unlike virtually all of the other key texts in this primer, *Geography and Gender* was written explicitly with undergraduates in mind; the writing is refreshingly clear and engaging. In view of the book's overall message and intent, however, it is clear that the intended audience extended well beyond undergraduates, to encompass the discipline as a whole. The main message is that it is important to understand how gender helps to structure geographies of inequality if that inequality is ever to be erased or at the very least significantly reduced. Juxtaposing gender (socially constructed difference) vs sex (biological difference), the authors understand gender relations to be malleable precisely because they are socially constructed and therefore subject to change. They recognize that such change – i.e., the kind of change in gender relations that will yield greater gender equality – will demand far more than simply making women visible in geography; it will require nothing less than 'an entirely different approach to geography as a whole' (WGSG, 1984: 21). In other words, these nine authors well understood the radical challenge that feminist geography posed to the entire discipline as well as to students.

The book joins an interest in explaining the value of a feminist approach in geography with a concern to make clear the tenuous position of women in UK geography. The discipline was decidedly masculine (only 10% of all full-time lecturing staff in the 50 UK geography departments for which they had survey data were women) and masculinist, in the sense that women were simply invisible in geography texts and other teaching materials. As I have argued elsewhere for the American case (Hanson, 2004), a look at the discipline's history reveals a close connection between the social identities of its practitioners and the topics deemed worthy of research attention, the research approaches used, the nature of data collected, and so on. *Geography and Gender* was the first book-length treatment of feminist geography, and it provided a glimpse not only of the gender of geography in the UK at the time but also of how geographical thinking might contribute to understandings of women's subordination.

The book is divided into three unequal parts. In the first, the authors lay out their arguments for studying feminist geography and describe four main approaches to understanding the unequal power relations that undergird gender-based inequality: radical feminism, socialist feminism, Marxist approaches, and phenomenological and humanistic approaches. The second part, which comprises the heart of the book, contains four chapters, each one focused on demonstrating the power of a feminist geographic analysis to produce new understandings in a core area of geographic knowledge: (1) urban spatial structure; (2) women's employment, industrial location, and regional change; (3) access to facilities; and (4) women and development. In the third part of the book the authors turn to the status of women in UK geography and the ways in which feminism should change approaches to teaching and research. Before considering the book's reception and impact, let us first look more closely at what these authors were saying.

In Part One, after making persuasive arguments about women's invisibility in geography and the need to overhaul geographic theory so as to understand why women remain in subordinate positions, the authors make clear their preference for socialist feminism over the alternatives described. Socialist feminists recognize that male family members and employers alike currently exploit women's unpaid work in the home, but they also believe that the state has a key role to play in effecting positive change for women, for example by making education and employment opportunities equal, by enabling women to live free of male violence, and by supporting services that allow women and men to combine family and work responsibilities. Feminists seek to change the world, and socialist feminists advocate women and men uniting to pressure the state to enact policies that support gender equality.

The four chapters in Part Two take up topics of longstanding interest to geographers and demonstrate how feminist insights challenge traditional theories. Each incorporates the relatively small amount of feminist research that had been done by the early 1980s, and the empirical material used in each chapter is decidedly UK-based. The chapter on *urban spatial structure* asks students to think about the implications of the spatial separation of home and work that accompanied industrialization. The authors trace out some of the historical ways in which the separation of home and work emerged from the division of labor between female domestic workers and male waged workers. After describing some of the progressive changes in living patterns initiated during World War II (e.g., housing with communal kitchens and child care), they describe the distinctly patriarchal dimensions of post-WWII urban planning and invite students to imagine what a non-sexist city might look like. The chapter on *women's employment, industrial location, and regional change* links changes in women's employment patterns (increased labor force participation, increased gender-based segregation in the labor market, and distinct geographic patterning in women's work) to larger processes of economic and regional change within the UK. The authors point out that a failure to disaggregate employment patterns by gender obscures some of the key processes that actually drive regional change, such as the distinct geographical variation in the demand for female labor of different types, a variation that was poorly understood at the time.

The third chapter in Part Two focuses on *access to facilities*. Through a case study of women's access to health facilities, the authors show that access – the availability of services – entails far more than just the ability to traverse space (travel); access also involves the hours of operation and the quality of the service at a facility *vis-à-vis* women's needs and constraints. *Women and development* is the focus of the fourth chapter in Part Two. Noting that, like other theories in geography, development theory has neglected women, the authors suggest that development policies affect women differently from men, that migration streams are gendered in different ways on different continents, and that young, unmarried women are being exploited by capitalists locating certain types of industries in developing countries.

Part Three, Feminism and Methods of Teaching and Research in Geography, documents the disadvantaged status of women in UK geography departments, raises questions about differential treatment of male and female students by university teachers (however inadvertent such treatment may be), offers ideas for student projects, and discusses the ways in which feminist ideas shape the research process. In this last vein, the authors have, throughout the substantive chapters of Part Two, made suggestions about the need for adopting feminist-inspired methodological approaches, such as supplementing quantitative surveys about health with in-depth interviews with women to discern the degree to which their health needs are being met or, in the context of understanding women's employment, listening to what women have to say about their everyday lives. Part Three also includes a plea to pay attention to the gender basis of male activities. The authors conclude their book by reminding readers that the goal of feminist geography is to change gender relations and that we can attain this goal only if we fundamentally change our theoretical and empirical approaches to research.

As I re-read the book recently, I was struck by how the core message – pay attention to gender if you really want to understand geographic processes – still resonates. Yet in 1984, when *Geography and Gender* first appeared, this message was revolutionary! The disciplinary norm was to ignore gender with abandon. Undoubtedly, this key text played a key role in creating some of the changes we have seen in geography over the past 20 years, e.g., the increased presence of women in academic geography, the recognition that gender thoroughly infuses geographic processes, and the now-rather-large and rapidly growing body of literature in feminist geography. Particularly as I re-read the problem-focused feminist analyses in Part Two on employment, access, and development, I was struck by how vastly much more we now know about these problems than we did in the early 1980s.

Although some of *Geography and Gender*'s central arguments do transcend time and place, the book is also very much a product of a particular time and a particular place. In this regard, I consider, first, some of the ways in which the authors' feminist arguments reflect the feminism of the time, and, second, some of the ways in which the book is very much rooted in the UK. The strong socialist feminist stance that infuses the text and the very concept of feminist geography espoused ('feminist geographers are concerned with the structure of social and spatial relations that contributes to women's oppression') signifies dominant concerns of feminism in the 1970s and early 1980s (WGSG, 1984: 134). The focus on uniting women as women, a group whose strongly shared gender identity was assumed, pre-dates concerns that arose in the mid-1980s about profound differences among women's experiences and positions and the problems such diversity poses for feminist theory and politics. The concept of patriarchy, which figures prominently in *Geography and Gender* and which was very much on the minds of feminist theorists at the time (see, for example, Foord and Gregson, 1986), rarely appears in contemporary feminist analyses; patriarchy as an explanatory framework in the 'grand theory' tradition has by-and-large fallen prey to post-structuralist views, which locate power in contextually variable discourses and practices rather than in grand structures.

The nine authors are all British, and their experiences, reflected in the theories and empirical examples selected for inclusion in *Geography and Gender*, were largely rooted in events and debates that were unfolding in the UK at the time. One might argue, too, that the particular approach to feminism they espoused was linked to their being situated in the urban UK. Certainly the topics that form the focus of the chapters in Part Two, while having wide currency then and now, reflect the authors' embeddedness within the UK and within the UK geography of the early 1980s. The chapter on women's employment, industrial location, and regional change, for example, draws women's employment into the then-raging debates about industrial restructuring and the massive economic disruptions that were accompanying that restructuring. Similarly, the chapter on access takes up a long-standing theme in human geography, one that, with its inherent concern for equity, was especially relevant in a UK undergoing profound social transformations.

We are all parochial, in the sense that we tend to write about what we know, and what we know is bound to the particular places and times in which we find ourselves. *Geography and Gender* was written with students, and mainly British students, in mind; although publishers are always asking textbook authors to write for a global student audience, the place from which an author writes inevitably pokes through to declare itself again and again throughout a text. This place declaration may be particularly striking to a non-local reader. As I was reading *Geography and Gender* with my American origins and perspective, I was often asking myself, 'why didn't they seem to know about *x*, or why didn't they mention *y*,' where *x* and *y* were publications by American feminist geographers available when this book was written. Examples are the six-article special feature on women, published in *The Professional Geographer* in 1982 and Mazey and Lee's (1983) *Her Space, Her Place*, also a book aimed at undergraduates.

Reception and evaluation

Upon its publication, *Geography and Gender* was welcomed as providing a much-needed introduction to feminist geography; the *fact* of the book – its very appearance – was heralded as much as was the contribution of its message. Those who reviewed the book were for the most part laudatory. Sallie Marston, reviewing for *Urban Resources*, a journal aimed at educational issues, praised the authors for attempting 'to lay out the foundation for a truly feminist geographic theory' (Marston, 1986: 60). In an appreciative review appearing in *Contemporary Issues in Geography and Education*, Linda Peake (1989) hailed the book's challenges to the traditional divisions within the discipline (i.e., those between the economic, social, political, and so on). Risa Palm (1986) in *Progress in Human*

Geography, and Ron Johnston (1985) in *British Book News*, while generally positive about the book's contributions, also found fault: Palm was critical of 'insufficient documentation for a number of claims' (Palm, 1986: 466), and Johnston complained that 'in places the geography is not as explicit as it could have been' (Johnston, 1985: 19). The initial reception within geography was on the whole, however, strongly supportive of these authors' efforts.

A thoughtful review by someone who was not familiar with feminist work in geography and was a self-proclaimed lapsed geographer, appeared in *Feminist Review* (Neligan, 1985). Although grateful to this group of feminist geographers for taking on the job of providing students with an introduction to feminist geography and appreciative of their challenges to geography traditions, Annie Neligan was disappointed in the (in her opinion) elementary level of the analysis. Of the discussion about zoning and urban spatial structure, she says, 'I wanted the writers to go on to discuss the effects of this zoning on women's lives and how these designs have perpetuated the divisions of labor that they reflect. They become one of the factors which confine women to part-time work, to taking low-paid service jobs that can be done near home. They add to the stresses of combining a job with collecting the children from school, so that home working becomes not an attractive but the only possible option' (Neligan, 1985: 115). She believes that in the section on women and development the authors have not adequately illuminated how 'international capital combine[s] with the patriarchy of traditional family life to produce the micro-chip assembly lines worked by young, Southeast Asian women, and how ... this feed[s] back into the location of investment' (Neligan, 1985: 115). I find these laments fascinating because in her articulation of what she finds missing in *Geography and Gender* Neligan is also putting her finger on just a small portion of what we

have learned about gender and geography since 1984.

Impacts and effects

The impacts of *Geography and Gender* can still be felt throughout the discipline, perhaps most palpably for students in geography curricula. In 1984, as now, there was considerable debate as to whether feminist geography should be taught in stand-alone courses or infused into courses throughout the curriculum. The authors of *Geography and Gender* advocated pursuit of both approaches, and my sense is that both have taken hold, not only in Anglophone countries but in many other countries as well. In the US, students have come to expect their instructors to incorporate feminist perspectives and gender into a range of human geography courses, and they complain if these are absent. Thanks in part to *Geography and Gender*, feminist analyses have become part of the geographical imagination that engage geography students worldwide; it is no longer acceptable to study space and place with gender absent.

Because *Geography and Gender* is rooted in the context of the UK in the early 1980s, it has sparked lively and sometimes acrimonious debate among feminist geographers, especially in Britain, about the appropriate scope and focus of feminist geography. In view of the depth of the book's challenges to traditional geography, such debate is entirely fitting. Indeed, one might argue that feminist geography, and *Geography and Gender* in articulating the cutting edge of feminist geography in 1984, was a key initiator of the broader critical assessment of geographical theory (especially its devotion to the status quo and resistance to change) that followed. Although the book's authors did not explicitly recognize the importance of diversity among women or the situatedness of a researcher, their work – by providing a coherent, cogent, and compelling extended argument for the necessity of feminist geography – laid the groundwork for the post-structural approaches that are now widely practiced in the discipline. This outcome is ironic in that some of the authors have resisted post-structuralism because of the difficulties it poses for political action. Nevertheless, a major reason for the lasting influence of *Geography and Gender* is, no doubt, that the authors went on to have distinguished careers in which they have continued to contribute to the development of feminist geography.

Indicative of this lasting influence, in 2001 *Geography and Gender* was the subject of a 'Classics in Human Geography Revisited' focus in *Progress in Human Geography*; brief commentaries by Robyn Longhurst and Susan Smith were accompanied by a collective response from the authors, in which they reflected on their 1984 book in light of subsequent developments. Both commentaries are laudatory, offering sympathetic critiques of what seems, from today's vantage point, to be oddly missing from the book.

Robyn Longhurst praises *Geography and Gender* as pathbreaking in three ways: providing a book-length treatment of gender in the field, undertaking a gendered analysis of several key topics in geography, and, in its collective authorship, challenging 'the politics of academic authorship' (Longhurst 2001: 253). She also calls the book visionary for foreshadowing several issues that have since been greatly illuminated by feminist geographers: work on men and masculinities; the need for feminist thinking to transcend disciplinary sub-fields, and the infusion of critical theorizing. As issues that, from a contemporary perspective, seem jarring, Longhurst identifies the authors' treatment of sex and gender as if they are separable and their decision to speak in one collective voice, as questionable given this ignores differences within feminism (in their response, the authors' rejoinder that they did recognize political differences among feminists in their description of the different approaches to feminism).

Susan Smith reflects on the book's contribution: 'while geographies without women still appear in the literature, the much greater sensitivity to gender and sexuality in the discipline's major journals is probably as good a marker as any of what this book helped achieve' (Smith, 2001: 255). She comments that, from today's vantage point, the emphasis in *Geography and Gender* on gender differences (and their social construction) to the neglect of differences/diversity among women seems like a shortcoming. Were they writing today, the authors would pay more attention to emotion, love, caring work and would recognize that gender is not a stable category but one that is performed differently in different contexts. Noting that, in 2000, only 6% of Professors of Geography in UK were women, Smith concludes, 'Things have moved on since 1984, but not so far that this classic text on *Geography and Gender* has lost its cutting edge' (Smith, 2001: 257).

In their response to these commentaries, the authors' assessment of their book 15 years after publication is instructive. Above all, they still believe that 'the book's central point, namely that understanding gender makes a difference to our geographical imaginations, remains as relevant today as it did then' (Smith, 2001: 257). Although the authors now recognize that their decision to write with a collective voice did suppress the diversity in their life experiences and subject positions, they note that the decision itself came from a desire to emphasize that all knowledge production is collaborative and to stress their common ground as women (along with one man) in geography. To them, the most significant shifts in feminist scholarship since the mid-1980s have been 'the epistemological and methodological implications of recognizing diversity' and the 'new theorization of sex and gender [that] have made space for researching and reconceptualizing the body' (Smith, 2001: 258). As they further explain regarding this last point, 'Issues about sexuality and embodiment [in the early 1980s] seemed only to cement women firmly into the "nature" side of the nature-culture binary' (Smith, 2001: 258). The authors are also now more circumspect about their ability to effect change, and this is the case for at least two reasons: the imposition of neoliberal policies and the difficulty of uniting people against gender oppression in the face of the many dimensions of diversity among women (and men).

In closing, the authors of *Geography and Gender* observe that, although more women are now present, academic geography in the UK is still dominantly white, male, and underpinned (perhaps now more than ever) by a masculinist, competitive ideology. They bemoan the absence of feminist principles in UK academic life. Although it seems fair to give these authors the last word on their own book, I would not end on such a pessimistic note. The very presence of more women and more feminists in geography, and increasingly in positions of power within geography and within individual academic institutions – not only in the UK but in countries around the world – signals hope for the kinds of positive change that the authors of *Geography and Gender* articulated so well in 1984.

Note

1 Sophie Bowlby, Jo Foord, Eleanore Kofman, Jane Lethbridge, Jane Lewis, Linda McDowell, Janet Momsen, John Silk, and Jacqueline Tivers.

Secondary sources and references

Foord, J. and Gregson, N. (1986) 'Patriarchy: Towards a reconceptualisation', *Antipode* 18 (2): 186–211.

Hanson, S. (2004) 'Who are 'we'? An important question for geography's future', *Annals of the Association of American Geographers* 94 (4): 715–722.

Johnston, R.J. (1985) 'Review of *Geography and Gender*', *British Book News*, January 1985: 18–19.

Longhurst, R. (2001) 'Commentary on *Geography and Gender*', *Progress in Human Geography* 25 (2): 253–255.

Marston, S. (1986) 'Review of Geography and Gender', *Urban Resources* 3 (2): 60–62.

Mazey, M.E. and Lee, D. (1983) *Her Space, Her Place: A Geography of Women.* Washington, D.C.: Association of American Geographers, Resource Publications in Geography.

Neligan, A. (1985) 'Review of 'Geography and Gender: An Introduction to Feminist Geography', *Feminist Review* 20: 113–118.

Palm, R. (1986) 'Review of *Geography and Gender*', *Progress in Human Geography* 10 (3): 466–467.

Peake, L. (1989) 'Review of *Geography and Gender*', *Contemporary Issues in Geography and Education: The Journal of the Association for Curriculum Development* 3 (1): 87–90.

Smith, S. (2001) 'Commentary on *Geography and Gender*', *Progress in Human Geography* 25 (2): 255–257.

Women and Geography Study Group of the IBG (1984) *Geography and Gender: An Introduction to Feminist Geography.* Harlow, Longmans Group.

12 SOCIAL FORMATION AND SYMBOLIC LANDSCAPE (1984): DENIS COSGROVE

David Gilbert

The Palladian country house and its enclosed parkland of sweeping lawns, artistically grouped trees and serpentine lakes offers a synthesis of motifs owing their origins to a range of sources: late renaissance Italy, classical humanism, the literary pastoral and the seventeenth- century painters in Rome. The finest of these 'landscapes', the parks at Stowe, Stourhead, Castle Howard, Chatsworth, Blenheim or The Leasowes, have come to be regarded as representative almost of the very character of the English countryside. From a rather different perspective they represent the victory of a new concept of landownership, best identified by that favourite eighteenth century word, *property*. The ideology of English parkland landscape may perhaps best be introduced by examining the design and iconography of one example, among the earliest and best-preserved of all, the garden at Rousham in Oxfordshire. (Cosgrove, 1984: 199)

Introduction

Rousham House stands about 12 miles north of Oxford in the English Midlands, on the west bank of the River Cherwell. The house was built in 1635, but Rousham is best known for its landscaped garden, designed by William Kent between 1737 and 1740. Although open to the public, visiting Rousham can be a little daunting. Unlike many other country houses, Rousham doesn't have a gift shop, nor does it have glossy posters or interactive displays about its history and design. No dogs and no children under 15 are admitted. The garden at Rousham provides one of the key examples in Denis Cosgrove's *Social Formation and Symbolic Landscape*, first published in 1984, and at first sight some students seem to find the book equally daunting. While the book is elegantly written, flicking through it can be somewhat intimidating. This is a book with a big subject – the history of the idea of landscape in the West – that, unlike most geographical texts, is unafraid to consider the long sweep of history. The illustrations, of fine-art landscape paintings from the Renaissance to the mid-twentieth century, or of the buildings and plans of the sixteenth-century Venetian architect Andrea Palladio, give an indication of the seemingly difficult materials (certainly unfamiliar to many geography students) that Cosgrove uses in the book. Like Rousham, *Social Formation and Symbolic Landscape* is a work that demands and repays serious consideration and detailed thought. And, like Rousham, it is also a work of carefully crafted details but with a clear overall design, a single central argument that has been influential in and beyond cultural geography.

Looking back, *Social Formation and Symbolic Landscape* is important because it established a new politicized notion of landscape as a fundamental concept in human geography. Cosgrove's approach to the idea of landscape is summarized in an often-quoted passage at the start of the book:

The landscape idea represents a way of seeing – a way in which some Europeans have represented to themselves and to others the world about them and their relationships with it, and through which they have commented on social relations. Landscape is a way of seeing that has its own history, but a history that can be understood only as part of a wider history of economy and society; that has its own assumptions and consequences, but assumptions and consequences whose origins and implications extend well beyond the use and perception of land; that has its own techniques of expression, but techniques which it shares with other areas of cultural practice. (Cosgrove, 1984: 1)

It is worth looking closely at this passage. Landscape is, as Cosgrove (1984: 13) puts it, 'an imprecise and ambiguous concept' without a single clearly agreed meaning. In everyday language, we may talk about landscapes in the physical world – 'the landscapes of New Zealand' – or we may think about landscape as a particular form of representation, often in art – 'the landscapes of John Constable'. Cosgrove's book is about both, as he argues that landscape art is one particular form of a more general 'way of seeing' the world.

In this context, the phrase, 'way of seeing', has a specific meaning. It refers to the approaches of Marxist art historians of the 1970s and 1980s, and specifically to the work of the British cultural critic, John Berger, who presented a BBC TV programme and wrote a subsequent book called *Ways of Seeing* (Berger, 1972) (though Berger himself took the terminology from the early twentieth-century art historian Erwin Panofsky). This approach argued that art could only be understood properly and fully in relation to wider contexts of social and economic power. Art is therefore not simply an individual creative activity, undertaken by lone artists. Who is able to produce art, what is defined as art, and how it is seen and used, are all questions that cannot be answered without reference to broader power relations. Indeed, this approach argued forcefully that the very category of art as a separate mode of human activity depended on particular forms of social and economic organization. Put most simply, the central argument of *Social Formation and Symbolic Landscape* is that this understanding of the relationship between social order and art is also directly applicable to the study of landscape.

The ordering of the two main terms in the title is an indication of the influence of Marxism on Cosgrove's thinking. 'Social Formation' is given precedence over the symbolism of landscape. Cosgrove was at pains to distance his work from approaches that treated 'the landscape way of seeing in a vacuum outside of the context of a real historical world of productive human relations' (Cosgrove, 1984: 2). However, if Cosgrove set himself against approaches to landscapes that interpret them with abstract references to their beauty, or as the work of individual geniuses, the terminology of the title also indicates his dissatisfaction with cruder varieties of Marxism. In some versions of Marxism, cultural products like literature, painting, poetry or drama were regarded as secondary, superficial aspects of human activity, to be understood as determined by the more fundamental economic organization of a society. In using the term 'social formation' rather than 'mode of production' which was more commonly associated with this deterministic Marxism, Cosgrove positioned himself alongside influential British Marxists of the period, such as the historian Edward Thompson and the cultural critic Raymond Williams. Such work rejected simplistic notions of an economic base and cultural superstructure, and argued culture was not a by-product of more fundamental conflicts between social classes over economic resources, but was 'an active force in the reproduction and change of social relations' (Cosgrove, 1984: 57).

Indeed, *Social Formation and Symbolic Landscape* can be seen as a development and extension of the arguments of one of Raymond Williams' best-known works, *The Country and the City*, published in 1973. Williams explored the representation of the rural and the urban in English literature, arguing both that literature reflected fundamental power structures, but also that culture worked as an active force in society that benefited the capitalist classes. Williams focused directly on the power of geography in culture, particularly on the significance of the representation of the countryside and the city; Cosgrove extended this to consider the landscape both as visual art and as a direct manipulation of the physical environment.

Reviewing the book in 1987, James Duncan argued that it was the book's sophisticated Marxist framework that gave it 'the potential to shift the direction of a subfield' (Duncan, 1987: 309). Later reviews have also highlighted the significance of the central argument. Don Mitchell, reflecting on the influence of the book on the development of cultural geography, stressed that without this strong argument situating the landscape way of seeing in the 'wider history of economy and society', the book would have been 'an accomplished and compelling synthetic description of changes in the nature of landscape representation … but would have done nothing to explain those changes' (Mitchell, 1999: 505). However, in his new introduction to the 1998 edition, Cosgrove stepped back from this strong explanatory framework, and following a shift made by many in Human Geography and beyond, exchanged his focus on capitalism and class relations for a more generalized and diffuse concern with the development of modernity. Perspectives such as feminism and post-colonial studies have broadened debates about the relations between

social power and cultural forms since the first publication of *Social Formation and Symbolic Landscape* (see Phillips, 2005; Pratt, 2005). In reading the book more than 20 years after its publication, we need to ask whether it is too bound into the intellectual culture and debates of its own period to have much relevance now, or whether it should be treated as an important starting point for later debates about landscape in particular and culture more generally.

The book and its author

Denis Cosgrove is probably best known as one of the main figures in what came to be known as the 'New Cultural Geography' (see Lilley, 2004 for more details of Cosgrove's career). Alongside others such as Peter Jackson and James Duncan, Cosgrove argued for a radicalized cultural geography, often in direct contra-distinction to an earlier established tradition associated with the work and legacy of the American cultural geographer, Carl Sauer (see Cosgrove, 1993; Cosgrove and Jackson, 1987). In *Social Formation and Symbolic Landscape*, Cosgrove did point to the importance of early twentieth-century cultural geographers such as Sauer and Paul Vidal de la Blache in developing perspectives that emphasized landscapes as material products of human societies. However, Cosgrove, Duncan, Jackson and others criticized the way that cultural geography, certainly as taught in universities in the US in the 1970s and 1980s, had reduced this legacy to a set of rather dull empirical techniques, often involving the detailed analysis of surviving 'traditional' landscapes (for discussions of the 'New Cultural Geography' and its relationship with Sauer's 'Berkeley School' see the debates in Price and Lewis, 1993a, 1993b; Cosgrove, 1993; Duncan, 1993; Jackson, 1993). The 'New Cultural Geography' combined Cosgrove and Duncan's focus on the politics of landscape with another strand of work that concentrated on the cultural politics and spatialities of identity. In the 1980s, the influence of Marxists like Raymond Williams remained strong; Peter Jackson's landmark 1989 textbook *Maps of Meaning*, while extending its scope beyond class to address the cultural geographies of race and gender, remained firmly grounded in Williams' cultural Marxism. By the 1990s, however, this new tradition had become much more fluid in its understanding of the nature of social power and its relations with culture – not least in the work of those academics most closely associated with the beginnings of the 'New Cultural Geography'.

Social Formation and Symbolic Landscape was Denis Cosgrove's first book, written while he was a Lecturer at Loughborough University. During his time in the East Midlands, Cosgrove developed a close intellectual relationship with Stephen Daniels of the University of Nottingham. The kind of materialist and political readings of landscapes and landscape art propounded in *Social Formation and Symbolic Landscape* became characteristic of an important strand of East Midlands Geography in the 1980s and 1990s. Perhaps the best-known work in this tradition is the collection co-edited by Cosgrove and Daniels in 1988, *The Iconography of Landscape*. The collection brought authors from art history, theology and literature together with historical geographers in studies ranging from sixteenth-century Italy to twentieth-century Canada, all concerned to decipher the social power of various forms of landscape imagery. Although the introductory essay is quite short, it develops significantly the arguments in *Social Formation and Symbolic Landscape*. There is a much stronger emphasis on the method required to interpret landscapes and landscape art, advocating Panofsky's 'deep iconography'. Pictures, and indeed actual landscapes, were to be treated as encoded 'texts', closely examined to reveal not just the meanings that had been consciously put there by an artist, landowner or landscape designer. This method of iconography was also a way of revealing intrinsic meanings about social power

structures that had shaped the making of the landscape.

However, beyond his influence on the discipline, Denis Cosgrove has also been that rather rare beast, a British geographer with significant standing across the humanities, and also beyond the English-speaking world. As such, while *Social Formation and Symbolic Landscape* is now regarded within Geography as a period piece, important in its time and significant for redirecting the sub-discipline of cultural geography, it has had a much more active and lasting presence in disciplines such as art history and landscape history, where it has retained currency for the significance of its central argument, but more for its detailed case-studies of the landscape idea in Italy, North America and England. Some chapters draw upon Cosgrove's doctoral work, a much more tightly focused project on the landscapes of post-renaissance Italy, particularly in Venice and the Veneto in the sixteenth century. This work culminated in the book *The Palladian Landscape,* published in 1993. Significantly, both this book and *Social Formation and Symbolic Landscape* have been translated into Italian.

Reading *Social Formation and Symbolic Landscape*

It is probably best to approach *Social Formation and Symbolic Landscape* through the commentary and reflections provided by Denis Cosgrove in his introduction to the 1998 edition. This is not a straightforward summary of the arguments, but places the book in the context of subsequent developments in cultural geography and landscape studies. *Social Formation and Symbolic Landscape* was also included recently in the 'classics in human geography' series in the journal *Progress in Human Geography*. Here Cosgrove replied to commentaries on the book by Lawrence Berg and James Duncan, summarizing some of the key points from his 1998 introduction (Berg *et al.*, 2005) A third route into the arguments of the book is to look at Cosgrove's article 'Prospect, perspective and the evolution of the landscape idea', published in 1985. This article overlaps with parts of the book, but concentrates on one aspect of its argument, considering how the 'invention' of modern techniques of perspective in the Italian renaissance contributed to the exercise of power over space through the landscape 'way of seeing'.

Social Formation and Symbolic Landscape can be treated as having two parts. The first part, consisting of the introduction and chapters on 'the idea of landscape' and 'landscape and social formation', is best thought of as a long essay expounding the theoretical themes of the book, while Chapters Three to Nine form detailed historical studies. The introduction itself is (unsurprisingly) the best place to start, but there are also sections that repay close attention in the second chapter. A key passage ('landscape and perspective', pp. 20–27) considers the power of perspective. In 1435, the Florentine architect Alberti published *Della Pittura* (or *On Painting*) in which he set out the principles of 'linear' or 'single point' perspective. These principles have formed the basis of what we might describe as a 'realist' representation of three-dimensional space, and were dominant in Western art between the renaissance and the late-nineteenth century. To our eyes early medieval paintings appear flat and distorted, while later paintings appear much more life-like or photographic in quality. These later paintings appear more real because they use converging lines to focus on a single point, the eye of the painter and the eye of the imagined observer.

Cosgrove connects Alberti's 'way of seeing' to the power structures of renaissance Italy, and particularly to the emergence of new forms of capitalist organization in the city-states of Tuscany. His argument is that artistic perspective, like parallel developments in accountancy, navigation, surveying, mapping and the science of artillery, helped form what he describes as the figure of the bourgeois individual, the powerful man placed at the centre of a world that he saw, owned and ordered. Cosgrove quotes Leonardo da Vinci's comment that the use of perspective 'transforms the mind of the painter into the likeness of the divine mind, for with a free hand he can produce different beings, animals, plants, fruits, landscapes, open fields, abysses and fearful places' (Cosgrove, 1985: 52). However, Cosgrove takes this further to suggest that in this distanced, often aerial view of landscape, sometimes described as the 'sovereign eye', it is ultimately not the individual painter that takes the role of God, but those who own and control the land.

In the remainder of his chapter on the 'idea of landscape', Cosgrove extends this argument about power and ways of seeing landscape to include modern geography as it developed as a discipline from the late-nineteenth century onwards: 'in some respects geography's concept of landscape may be regarded as the formalising of a world-view first developed in painting and the arts into a systematic body of knowledge claiming the validity of a science' (Cosgrove, 1984: 27). In a key section that has strong resonances for current debates in the discipline about representation (see Rose *et al.*, 2003), Cosgrove argues that Geography has suffered from a 'visual bias' which has meant that it has been obsessed with surveying, mapping and representing landscapes, rather than interrogating them as the constructions of particular social formations. He praises the Humanistic Geography of the 1970s and early 1980s for emphasizing the emotions and sensations of human beings living within landscapes, but argues that this cannot be enough. The task for the critical geographer is to 'trace the history of the landscape way of seeing and controlling the world' (Cosgrove, 1984: 38).

The following chapter ('Landscape and Social Formation: Theoretical Considerations') starts this task with a very broad discussion of the long-term history of social and economic development in the West. It's a difficult chapter, which centres on contemporary debates about the transition from feudalism to capitalism. The key points are: firstly, that this transition in the fundamental economic, social and political organization of Western societies was a long-term process; secondly, that it took place at different rates and in different forms in different countries; and thirdly that struggles over the ownership and control of land were central to the transition. These ideas are rather more easily understood in the historical case studies that follow in the second part of the book. These chapters consider, in turn, the relationships between social formations and landscape in renaissance Italy, in sixteenth-century Venice and the Veneto region, and in North America and England during the seventeenth and eighteenth centuries. Two final chapters explore the landscape way of seeing in the nineteenth and twentieth centuries. Each chapter has a broadly similar form, discussing first the dominant forms of landscape art, architecture and design in the period, before drawing out their basis in broader social conflicts and their use within those struggles.

It's useful to go back to Rousham for an example of this method in action. Cosgrove gives a detailed description of the landscape garden created by William Kent by the Cherwell (Cosgrove 1984: 199–206). Like more conventional art historical accounts of the garden, Cosgrove identifies the influence of painting on Kent's work, particularly the work of Claude Lorrain and Nicholas Poussin. Kent created a 'Claudian landscape' in 'the reality of lawns, trees and controlled views within and beyond the garden' (Cosgrove, 1984: 201). Kent's garden was designed to be seen and experienced in a carefully constructed circuit that combined details such as statues and ornamental bridges in the garden with views out into the English countryside

beyond. However, as well as examining the details of the garden, Cosgrove also interprets the designs at Rousham as part of wider developments in the cultural politics of landscape. The English landowning class used art and landscape design, together with certain literary forms, to express its position and power. The phrase 'commanding view' is often used loosely today to indicate a wide panorama from a high point in the landscape; Cosgrove shows that the constructed views of the great English parks were literally both commanding and commanded.

However, Cosgrove goes further than this rather blunt assertion that Rousham was a landscape of power, and much of the lasting value of the case study comes from the more nuanced way that he places the garden in the precise context of the cultural politics of the period. Rousham's constructed 'natural' landscape, with its irregular shapes and serpentine lines, was a direct challenge to earlier formal and geometric landscape gardening as at Versailles, and as such could be treated as a celebration of English bourgeois 'liberty' as against continental absolutism. The garden also combined classical elements (filtered through Palladian landscape conventions developed in the sixteenth-century Veneto) with views of a supposedly unadorned English countryside. This combination validated the taste and sensibility of the landowner, but also placed the beauty of the English landscape within a longer narrative of the rise of Western civilization. What Cosgrove achieves is a deep reading of the landscape at Rousham that gives due attention to artistic creativity and cultural traditions, but also to the economic and social milieu within which they worked.

Critiquing *Social Formation and Symbolic Landscape*

All of the historical chapters of *Social Formation and Symbolic Landscape* combine subtle readings of the politics of landscape with an underlying Marxian framework that gives explanatory priority to changing class relations. While the detailed case studies retain considerable force and interest, criticisms of the book, including Denis Cosgrove's own reflections, concern this underlying argument. At one level, this is part of broader developments in human geography and in social theory more generally that have challenged Marxist frameworks of understanding, whether in their more overtly determinist forms, or the more flexible, humanistic tradition underpinning *Social Formation and Symbolic Landscape*. Rather than rehearse these general critiques here, it is better to think about how they translated into more specific challenges to the arguments put forward by Cosgrove in 1984.

These challenges question the ultimate priority given to class in *Social Formation and Symbolic Landscape*. Subsequent work, particularly that drawing on feminist and post-colonial perspectives, has argued that a landscape way of seeing cannot be understood without references to other dimensions of social power, particularly those associated with gender and race. This challenge was made forcefully by Gillian Rose in *Feminism and Geography* (Rose, 1993: particularly 86–112). One of Cosgrove's strongest claims is that landscape is a 'visual ideology', taking a specific and partial world-view, and presenting it as a natural and necessary way of seeing. This class-bound view of the world erased the actual economic exploitation that made it possible, particularly of the rural working class. Rose extends this argument to claim not only that a similar erasure of the inequalities of gender takes place in the landscape way of seeing, but also that Cosgrove's account itself reinforces this through the almost complete absence of consideration of gender in the text. As Cosgrove acknowledges in his introduction to the 1998 edition, the artists and landowners discussed in *Social Formation and Symbolic Landscape* are not just all male, but also 'they appear and communicate to us as *eyes*, largely

disconnected from any other corporeal or sensual aspects of their being' (Cosgrove, 1998: xviii). Put another way, while all of Cosgrove's intellectual effort goes into revealing how the unacknowledged power of social class structured the work of these men, he has nothing to say about the impact of their masculinity.

This is a serious omission, openly recognized by Cosgrove in 1998. Although *Social Formation and Symbolic Landscape* was written before the expansion of feminist thinking that took place in Geography in the late-1980s, the importance of gender relations in 'ways of seeing' was already an important area of debate in the social sciences and humanities more generally. Berger's *Ways of Seeing* (1972) included a substantial section on the power relations in the visual representation of women by and for men. At about the same time, Laura Mulvey published her classic essay on the power of the 'male gaze' in cinema (Mulvey, 1975). Even working within the confines of a Marxist approach, such issues could and should have been woven into the argument of *Social Formation and Symbolic Landscape*, as the property relations discussed were clearly gendered. Gillian Rose, however, takes her critique further, arguing not merely that *Social Formation and Symbolic Landscape* ignores the material exploitation of women and their exclusion from the making of the landscape way of seeing, but also that Cosgrove fundamentally misreads the idea of the sovereign eye. For Rose, the central feature of this view of landscape is that it is a dominant masculine gaze. Cosgrove in response is more cautious about this critique, resisting the claim that 'the landscape idea inevitably constructs gendered landscapes as the passive, feminized objects of a rapacious and voyeuristic male gaze' while acknowledging the need for more attention being devoted to the role of sexual desire in landscape representation (Cosgrove, 1998: xviii).

A second criticism of *Social Formation and Symbolic Landscape* is also connected to its central emphasis on class relations and property, and relates to post-colonial theory. While the book gives close attention to the significance of the 'New World' of North America in the development of the European landscape idea, there is no consideration of the way that Europe's position within wider economic, political and cultural networks shaped ideas about landscape. This is in part about another form of erasure; for example, the great English estates discussed in the book were as much the product of imperial expansion and slavery abroad as they were of class power within England. *Social Formation and Symbolic Landscape* does not consider the export and hybridization of European landscape ideas in other environments as a part of the imperial project. And the book also fails to consider the ways that European landscapes were altered by the flow into Europe of new images and knowledge of 'exotic' environments, and particularly by the cultivation in Europe of plants from beyond its boundaries.

A final possible criticism of *Social Formation and Symbolic Landscape* is emphasized by Cosgrove himself (Cosgrove, 1998: xx–xxi; Cosgrove and Duncan, 2005: 481–482). The overall narrative of capitalist development used as the basis for the book suggests that by the twentieth century land and land-ownership had ceased to be a central arena of social and economic conflict. As a consequence, landscape was seen as of waning significance both in art, and more generally as a political and moral issue. Cosgrove's turn from this particular reading of the history of capitalism towards a more diffuse and complex concern for the making of the modern world calls this claim into question at a theoretical level. But, as Cosgrove suggests, this argument was also a misreading of the significance of landscape representation in modern times. By focusing on some schools of landscape art in the mid-twentieth century, he was able to argue that this was a form in decline, an increasingly irrelevant backwater in the development of modern art. This seriously underplayed the importance of landscape art in the development of the modern nation-state, which has subsequently been an important focus for cultural geographers (see Daniels, 1993; Matless, 1998). Furthermore, while Cosgrove recognized the work in the 1970s and 1980s of some landscape artists like Richard Long, he did not foresee the emergence of new kinds of landscape art that have disrupted the traditional form of the gaze, and which crucially have responded to new forms of environmental and identity politics (Nash, 1996).

Conclusion

Social Formation and Symbolic Landscape remains an important read today not so much because of its importance in the development of cultural geography, but because its central theme – the power of landscape and the landscape way of seeing – still matters. Few geographers, including Cosgrove himself, now accept the particular formulation of social power presented in the book, and its omissions are serious (although the way that it opened up, for example, discussion of the 'sovereign eye' in Geography made it easier for others with feminist or post-colonial perspectives to develop the argument). Despite its overwhelming concern with landownership and class relations, *Social Formation and Symbolic Landscape* was subtle enough to include tantalizing passages about the significance of embodied experience and of the impact of myths and collective memory, both of which have been important themes in recent cultural geography. The challenge inspired by reading *Social Formation and Symbolic Landscape* today is to imagine a new formulation that connects the wider history of economies, societies, nations and empires to the study of landscape, as a way both of seeing the world and of being in the world.

Secondary sources and references

Berg, L., Cosgrove, D. and Duncan, J. (2005) 'Commentaries and author's response in "Classics in human geography revisited": *Social Formation and Symbolic Landscape*,' *Progress in Human Geography* 29: 475–482.

Berger, J. (1972) *Ways of Seeing*. London: Penguin.

Cosgrove, D. (1983) 'Towards a radical cultural geography: problems of theory', *Antipode* 15: 1–11.

Cosgrove, D. (1984) *Social Formation and Symbolic Landscape*. Beckenham, Kent: Croom Helm.

Cosgrove, D. (1985) 'Prospect, perspective and the evolution of the landscape ideal', *Transactions of the Institute of British Geographers*, New Series, 10: 45–62.

Cosgrove, D. (1993) 'On "the reinvention of cultural geography" by Price and Lewis', *Annals of the Association of American Geographers* 83: 515–517.

Cosgrove, D. (1994) 'Contested global visions: One-World, Whole-Earth, and the Apollo Space Photographs', *Annals of the Association of American Geographers* 84: 270–294.

Cosgrove, D. (1998) *Social Formation and Symbolic Landscape* (second edition) with new introduction. Madison, WI: University of Wisconsin Press.

Cosgrove, D. and Daniels, S. (1988) *The Iconography of Landscape*. Cambridge: Cambridge University Press.

Cosgrove, D. and Jackson, P. (1987), 'New directions in cultural geography', *Area* 9: 95–101.

Daniels, S. (1993) *Fields of Vision: Landscape Imagery and National Identity in England and the United States*. Princeton: Princeton University Press.

Duncan, J. (1987) Review of Social Formation and Symbolic Landscape', *Annals of the Association of American Geographers* 77: 309–311.

Duncan, J. (1993) 'On "the reinvention of cultural geography" by Price and Lewis', *Annals of the Association of American Geographers* 83: 517–519.

Jackson, P. (1989) *Maps of Meaning: An Introduction to Cultural Geography*. London: Unwin Hyman.

Jackson, P. (1993) 'On "the reinvention of cultural geography" by Price and Lewis', *Annals of the Association of American Geographers* 83: 519–520.

Lilley, K. (2004) 'Denis Cosgrove', in P. Hubbard, R. Kitchin and G. Valentine (eds), *Key Thinkers on Space and Place*. London: Sage.

Matless, D. (1998) *Landscape and Englishness*. London: Reaktion.

Mitchell, D. (1999) Review of *Social Formation and Symbolic Landscape* (second edition). *Transactions of the Institute of British Geographers* 24: 505–506.

Mulvey, L. (1975) 'Visual pleasure and narrative cinema', *Screen* 16 (3): 6–18.

Nash, C. (1996) 'Reclaiming vision: Looking at landscape and the body', *Gender, Place and Culture* 3: 149–169.

Phillips, R. (2005) 'Colonialism and postcolonialism', in P. Cloke, P. Crang and M. Goodwin (eds), *Introducing Human Geographies*. London: Hodder Arnold.

Pratt, G. (2005) 'Masculinity-femininity', in P. Cloke, P. Crang and M. Goodwin (eds), *Introducing Human Geographies*. London: Hodder Arnold.

Price, M. and Lewis, M. (1993a) 'The reinvention of cultural geography', *Annals of the Association of American Geographers* 83: 1–17.

Price, M. and Lewis, M. (1993b) 'Reply: On reading cultural geography', *Annals of the Association of American Geographers* 83: 520–522.

Rose, G. (1993) *Feminism and Geography: The Limits of Geographical Knowledge.* Cambridge: Polity.

Rose, G., Matless, D., Driver, F., Ryan, J. and Crang, M. (2003) 'Intervention roundtable: Geographical knowledge and visual practices', *Antipode* 35: 212–243.

Williams, R. (1973) *The Country and the City.* London: Chatto & Windus.

13 CAPITALIST WORLD DEVELOPMENT (1986): STUART CORBRIDGE

Satish Kumar

Accounts of development and underdevelopment must be sensitive not only to the dynamic of the world system, but also to the relations of production and their conditions of existence that are the characteristics of particular countries and development sectors. (Corbridge, 1986: 154)

Introduction

Geographers have long documented how processes of uneven development have served to separate the haves from the have-nots. Global capitalism and neo-liberal reforms have worked to strengthen such divisions. In *Capitalist World Development*, Stuart Corbridge, a geographer and a development studies specialist who has worked extensively in India, sought to document the spatial imprints of uneven development and also provide a critique of radical geographical theories that sought to explain such processes. As such, the book needs to be understood as a product of a particular time, as a reaction to the Marxist theories of development that dominated geographical scholarship and the social sciences at the time of its writing. Corbridge thus wrote the book as a critique of radical development geography and its emphasis on epistemology and ideology rather than real-world experiences. In such Marxist accounts, the State is reified and economistic explanations only help to essentialize conclusions about the developing world. He calls for a dialogue between Marxists and non-Marxists to provide a realistic assessment of underdevelopment and provides a clear exposition of the intellectual origins and arguments of key radical developmental theories. In so doing, he astutely clarifies complex arguments and concepts. The result is a book that is neither pro-capitalist nor anti-Left. It is an attempt to go beyond trading dogmas and determinism in explaining the persistence of underdevelopment in the world, providing a focused introduction to the world of radical development geography.

Radical geography emerged as a reaction to growing social inequalities in the West, including concern over environmental issues, racism and global inequality. There was growing unease with Western foreign policy, particularly the Vietnam fiasco, and concern for the plight of the developing world. While some radical geographers shied away from Marxist theory, others actively engaged with its ideology and doctrine. Throughout the 1970s and early 1980s debate continued about the form and nature that radical geography should take – for example, scientific versus critical Marxism, theoretical or historical materialism, and structuralism versus humanism (see Duncan and Ley, 1982, for an extended analysis). In time, a materialist ontology came to dominate Marxist geography, one in which capitalism came to be viewed as the source of all worldly evils – not least inequality and unevenness of development. This marriage of Marxism and space produced exciting and compelling explanations of global injustice by highlighting how the persistence of uneven development is as important to capitalism as the direct exploitation of labour (see Castree, Chapter 8 this volume; Phelps, Chapter 10 this volume).

The book's key arguments

Capitalist World Development marked an attempt to challenge this dominant Marxist mode of an analysis, building on and extending critiques such as those developed by Slater (1976) and deSouza and Porter (1976). These authors, like Corbridge, asserted that neo-Marxist geography as then formulated was incapable of resolving developmental problems and by the middle of the 1980s scholars like Booth (1985) and Corbridge were attempting to understand the nature of *impasse* in Marxist approaches to development and underdevelopment. In particular, *Capitalist World Development* critiqued the dogmatism and essentialism associated with Marxist development theories, such as the primacy of 'structures' and the transcendental qualities of 'capital', calling for a change of the *'explanadum'* (that which requires explanation) to provide a solution to the *impasse* in development studies. In doing so it sought to move beyond the limitations of neo-Marxist and Marxists approaches without necessarily condemning their transformatory ambitions. Corbridge showed how questions of debt and demography are influenced by not just Marxist theory and ideology, but also with the material reality of the Third World.

Capitalist World Development starts with two basic questions, namely, 'What are the consequences of adopting a radical perspective on differential development?' and 'What are the consequences of adopting a view of events which emphasize a necessary conflict of interest between metropolitan capitalism and the development of the periphery?' The

answers were spelt out by Corbridge as follows:

- Capitalism cannot or will not promote development of the Third World.
- Capitalism alone is ultimately responsible for the world's demographic and environmental ills.
- Capitalism is incapable of promoting independent industrialization of the South.
- The fundamental divide shaping the capitalist world system is between the North and South.

In providing these answers Corbridge shifted the focus away from structuralism to emphasize the human face of structural constraints in world capitalist development. As such, overcoming the tendency to oppositionism, determinism, spatial over-aggregation, and epistemological confrontation was critical to Corbridge (1986: 9–13) in setting in motion a new agenda for development studies.

Oppositionism relates to a confrontational approach to a given argument, which becomes counter-productive when it refuses to acknowledge logical and relevant arguments. For example, the term 'capitalism' has been radicalized to provide a negative connotation, namely that of promoting underdevelopment in the Third World. This is all the more true when we evaluate the perspective that population growth rates are irrelevant to patterns of differential development. In fact, population growth may not necessarily cause underdevelopment (an argument illustrated by the growth of the economies of India and China in the twenty-first century). This can also be substantiated when we address the deterministic claim of radical development geography that capitalism is incapable of promoting industrialization of the Third World. Current world developments do not bear out this proposition. Thus Corbridge's (1986) claim that capitalism is capable of promoting industrialization of the periphery is particularly relevant today, especially when we see the emergence of the newly industrialized economies (NIEs) of Singapore, Malaysia, South Korea, Taiwan and Hong Kong and newly industrializing countries (NICs) of China and India.

Determinism was identified by Corbridge as another major problem in the radical approach to development geography. This relates to capitalist models of development and underdevelopment. Here capitalism can only promote underdevelopment in the periphery as it is largely based on an exploitation of both productive and unproductive resources. The suggested way to overcome this exploitation, according to the Marxian analysis, is to disengage economically with the developed world.

The third critique developed by Corbridge was that of spatial over-aggregation whereby the capitalist system is conceived as being one of north versus south or core versus periphery. Such forms of over-aggregation have tended to normalize and ignore the variety of capitalist forms of development and underdevelopment. The different shades of capitalism visible today relate to the diverse forms that neo-liberalism has articulated itself in space. Indeed, Corbridge's critique anticipates and predates the new ideas about the constantly transforming forms of capitalism current in the literature (see Peck and Tickell, 2002).

The fourth critique concerned the *epistemological confrontation* resorted to by radical development geographers. The lack of engagement between Marxists and non-Marxists has arguably resulted in a 'dialogue of the deaf' (Chisholm, 1982: 11; Corbridge, 1986: 12). Here, there is a clear disjuncture between theory, ideology and actual practice. The privileging and over-emphasis on economistic explanation reinforced essentialist and dogmatic conclusions, which were far removed from ground reality. Such a position has resulted in the failure of the Marxists and neo-Marxists to acknowledge the negative role that any given State can play in preventing development, particularly through corruption and reinforcing elitism. They have also ignored the potential of endogenous growth in the periphery. Corbridge asserts that dismissing an argument as less important for not subscribing to the radical viewpoint does disservice to academic engagement. The book emphatically states that a non-progressive view of capitalism dominated radical geography and a 'dialogue of the deaf' is of no use to the poor in the developing world whose future depends on the policies guided and informed by theories of development and underdevelopment (Corbridge, 1986: 13). By evaluating the role of capitalism in fostering development and underdevelopment, Corbridge offered a way out of the deterministic mode of thinking and suggests that there are no general laws of capitalist development which can become the basis for explaining the persistence of differential uneven development in the world. Accordingly, the central question should be how far capitalism can be held responsible for underdevelopment in the world?

To support his argument, Corbridge provides a brilliant critique of Harvey's (1974) paper on 'Population, resources and the ideology of science'. While accepting Harvey's incisive critique of Malthusian and evolutionary ideas of population growth, Corbridge argues Harvey's relentless epistemological dogmatism did not develop non-Malthusian perspectives on the population-development debates. Failure to engage with practical population problems and their politics of control in the developing world are clearly important and, according to Corbridge, were not acknowledged by the radical development geographers (Corbridge, 1986: 87–103).

The discussion on neo-Malthusian issues is still topical and relevant today, particularly in the context of sustainable development. The recent emergence of India and China as emerging world economic powers puts to rest all speculative ideas and ideologies of *catastrophism* inherent in Malthusian accounts. Indeed as Corbridge notes, 'rapid population growth at times exerts a secular influence upon economic growth' (1986: 84). This is confirmed by recent analyses emerging from South Asia, which overturn the old adage that 'the potency of population growth is a negative factor in India's economic performance' (see Corbridge, 1986: 90). The argument here is that demographic transition in selective regions actually aided the resurgence of economic growth in India. Despite the population pressure, India's economic growth has been consistently based on specific structural adjustments and liberalization of the economy. Thus, Corbridge's (1986) question whether it is possible (and/or desirable) to curb population growth to promote a wider restructuring of the economy is answered in the affirmative in the case of China and India. However, what is important to bear in mind is that both these countries have an established population momentum and this does not come to a halt overnight. So, despite the one child per family policy in China and demographic transition in southern Indian states, population growth has not come to a grinding halt. Therefore, Corbridge's (1986: 100) assertion that 'taken together these accounts leave us in little doubt that a decline in fertility can be secured in advance of radical structural changes, as the moderniser claims', is generally vindicated by current experiences.

The issue of climate change, also discussed in the book, has also become more topical now than ever and Corbridge refocuses the terms of debate across both the *Left* and *Right* of the political spectrum. The idea that the origins of environmental maladies are traceable to the inefficiencies of capitalism is a deeply contested argument, riddled with ideological rhetoric, and remains a potent source for debate among the radicals. Globalization has advanced a replication of production forms and exchanges between the developed and the developing world, thereby overturning the notion that less advanced nations are incapable of securing any lasting competitive advantage in the world of manufacturing,

services and trade. Indeed, while the existence of capitalism is not disputed, the TNC-led industrialization espoused by the neo-Marxists for the developing world is *passé* in current globalized context. Today the terms of reference have shifted from the *dirigismic* state to an entrepreneurial state and finally to a 'paternal state'. Notions that industrialization in the Third World is incompatible with capitalism (Corbridge, 1986: 131) now appear very dated, given nation states are vying with emergent markets to seek a fairer share of the global pie.

The book makes an important assertion that ideas of stagnation had no roots in Marxism and, in this context, the claims made by radical geographers in the context of the periphery appear rather dubious (Peet, 1975; Blaut, 1974). The persistence of underdevelopment in the periphery can therefore be apportioned to the internal contradictions in the Third World, such as policy blunders and fiscal mismanagement, as much due to structural constraints as to development. In line with this thought the moral critique of capitalism needs to be retained, particularly in the context of poverty and inequality in the Third World (Kumar, 2004).

In making such arguments, Corbridge provided a strong critique of neo-Marxists' lack of appreciation of the role of government in the Third World in assisting capitalist development. In a way the conclusion that Corbridge states still holds true, namely that 'accounts of development and underdevelopment must be sensitive not only to the dynamics of the world system, but also to the relations of production and their conditions of existence that are the characteristics of particular countries and development sectors' (Corbridge, 1986: 154). New forms of unequal exchange and asymmetries of power are being orchestrated despite the establishment of WTO. The global economic system remains embedded, thereby perpetuating inequality and disparity. In a sense, Corbridge's critique anticipated contemporary calls to analyse issues of neo-liberalism, which was more contextual and relational in its form and functioning, rather than a monolithic concept. He emphasized nation states as intrinsically empowered to determine the course of development within their boundaries. This of course is subject to qualification, in that functional democracy makes this possible wherever accountability is devolved into the hands of the citizens.

Critiques and legacy

Some of the critiques of Corbridge's work seemed largely guided by prejudice rather than factual argument (Booth, 1993). Indeed, some geographers who have reviewed this book have been highly critical (see *Progress in Human Geography*, 2005). For instance, virulent outbursts labelling the book as 'centre-right wing polemic writing in praise of monopoly capitalism and against radicals' (Blaut, 1989: 102) did not necessarily offer a fruitful engagement with the book's argument, but certainly exposed the deep sense of insecurity in radical scholarship. Blaut states 'Corbridge adds little to this standard argument of conservative social science … Radical geography in fact rather badly needs constructive criticism, but you won't find it in this volume'. Peet (1988: 190) stated, 'in brief, radical development geography is not criticised in terms of its own content but indirectly via various radical social theories whose connection with geography is unproven and largely unexamined'. Further, he adds, 'This is not a critique of radical development geography, which occupies no more than five percent of the text. On the few occasions geography is mentioned, simple caricatures are drawn, which reveal more about Corbridge than their objects' (Peet, 1988: 191). Again, Johns (1990: 180) states, 'the book's title is a bit misleading. The text fails to engage either the literature or the reality of *Capitalist World Development*; nor is it a sustained critique of radical development geography'.

Gaile (1988), on the other hand, stated [t]his book is pro-scholarship and is not pro-capitalist'. Likewise Brookfield (1987: 119) notes, 'that modern currents in non-Marxist development writings are inadequately treated … [but] the writing style is very lucid and the author has taken great care in his efforts to clarify complex and sometimes turgid arguments for his readers'. In presenting a detailed and critical analysis, the book reflects the extent of Corbridge's empathy with the developing world.

While the book never got its full dues on publication, it is clear that it helped move development studies beyond its *impasse* by outlining key themes in post-Marxist theory in the wake of *Perestroika* and its related *Glasnost* in the Soviet Union. Corbridge has since engaged with issues of world debt, the power and role of the state (see Corbridge and Harriss, 2000, *Reinventing India*, and Corbridge et al., 2005. *Seeing the State*). He has hence provided a coherent and important contribution to the critical debates of world capitalist development which is contingent and reflective of real issues in the developing world. Thus, in the current context his book may appear eclectic and critical of his contemporaries, but it cannot be faulted for being reductionist and dogmatic.

For a current readership it will be more appropriate to test his ideas in the wake of the recent development and transformations in the peripheral economies such as India and China. One has also to bear in mind that capitalism is constantly evolving and changing to suit the imperatives of world development. This book does much to translate difficult and obtuse ideas and concepts for non-specialists. It sets the terms of critical engagement, which has been followed by others in the radical genre. Corbridge asserts that one needs to go beyond dogmas and determinism to deconstruct the view that capitalism and its laws of motion are not fixed in time and space. Here, the exigencies of time and space are constantly mediated in the peripheries by conditions of population, economic growth, state policies, governance, etc. and therefore are not controlled by the inexorable tide of the 'grand world system'. The book accordingly highlights the internal incoherence of many of the radical theories adapted by radical development geographers such as Blaut (1970, 1974), Santos (1974), Slater (1976), and Peet (1975), to suggest an epistemological denouncement of development geographers is tantamount to throwing the baby out with the bath water and at no point does Corbridge suggest such a course of action. As far as is discernible, he calls for a realistic revision of the rhetoric that ideology supersedes theory based on evidence from the field. Indeed 'a radical perspective need not be so pessimistic' (Corbridge, 1993: 213).

The emerging consensus regarding development in the periphery from both the non-Marxists and those on the 'Left' is that traditional ideas of capitalism suggest the spontaneous diffusion of development from core to the peripheries (see Baran's thesis, 1957), yet what we observe today is not a spontaneous diffusion but a protracted struggle, enforced by the post-colonial state through instruments of protectionism, import substitution industrialization and the gradual dismantling of the *'dirigisme'* phase of development. The case in point is that diffusion of industrial development from the core to the periphery has not meant an all-round general development of the people in the developing world. Indeed, the very sustainability and effectiveness of these forms of development is highly suspect (Patnaik, 2006).

Conclusion

Regardless of reviewer responses, *Capitalist World Development* has become a standard reference in teaching modules dealing with issues of development across the spectrum of social sciences. Corbridge's critical insights present a necessary counterbalance to the perspectives provided by radical development theorists. The relevance that the book might have in the future is difficult to predict. Suffice to say that capitalism as an entity has not dissipated over time, neither has underdevelopment or poverty disappeared over the edges of global-local horizons. As long as we ponder and grapple with the crisis of accumulation in both the core and peripheries, as long as we engage with the impossibility of a singular approach to capitalism, of the *impasse* in development studies, *Capitalist World Development* will remind us of the need to avoid the 'dialogue of the deaf' at all times.

Secondary sources and references

Baran, P.A. (1957) *The Political Economy of Growth*. New York: Monthly Review Press.

Blaut, J. (1970) 'Geographic models of imperialism', *Antipode* 2: 65–85.

Blaut, J. (1974) 'The ghetto as an internal neo-colony', *Antipode* 6: 37–41.

Blaut, J. (1989) 'Review of Capitalist World Development: A critique of radical development geography', *The Professional Geographer* 41: 102–103.

Booth, D. (1985) 'Marxism and development sociology: Interpreting the impasse', *World Development* 13 (7): 761–787.

Booth, D. (1993) 'Development research: From impasse to a new agenda', in F. Schuurman (ed.) *Beyond the Impasse: New Directions in Development Theory*. London: Zed Books, pp. 49–76.

Brookfield, H. (1987) 'Review of Capitalist World Development: A critique of radical development geography', *The Journal of Development Studies* 24 (1): 118–119.

Chisholm, M. (1982) *Modern World Development: A Geographical Perspective.* London: Hutchinson.

Corbridge, S. (1986) *Capitalist World Development: A Critique of Radical Development Geography.* London: Macmillan.

Corbridge, S. (1993) 'Marxisms, modernities and moralities: development praxis and the claims of distant strangers', *Environment & Planning D: Society and Space* 11: 449–472.

Corbridge, S. and Harriss, J. (2000) *Reinventing India: Liberalisation, Hindu Nationalism and Popular Democracy.* Cambridge: Polity Press.

Corbridge, S. (2005) 'Classics in human geography revisited, Capitalist world development: a critique of radical development geography', *Progress in Human Geography,* 29 (5): 601–608. London: Macmillan.

Corbridge, S., Williams, G., Srivastava, M. and Veron, R. (2005) *Seeing the State: Governance and Governmentality in India.* Cambridge: Cambridge University Press.

deSouza, A.R. and Porter, P.W. (1976) 'Development geography and radical–liberal dialogue', *Antipode* 8 (3): 94–102.

Duncan, J. and Ley, D. (1982) 'Structural Marxism and human geography: A critical assessment', *Annals of the Association of American Geographers* 72 (1): 30–59.

Gaile, G.L. (1988) 'Review of capitalist world development: A critique of radical development geography' *Political Geography Quarterly* 7 (4): 373–377.

Harvey, D. (1974) 'Population, resources, and the ideology of science', *Economic Geography* 50: 256–277.

Johns, M. (1990) 'Review of Capitalist World Development: A critique of radical development geography', *Antipode* 22 (3): 177–180.

Kumar, M.S. (2004) 'Rhetoric of globalisation and reforms: A case of India (1991–2003)', in C.G. Krishnaswamy, S.R. Keshava and R.M. Tirlapur (eds), *Better Expression on Globalisation.* Bangalore: Bangalore University Press, pp. 23–52.

Patnaik, P. (2006) 'Diffusion of development', *Economic and Political Weekly,* May 6: 1766–1772.

Peck, J. and Tickell, A. (2002) 'Neoliberalising space', *Antipode* 34 (3): 380–404.

Peet, R. (1975) 'Inequality and poverty: A Marxist-geographic theory', *Annals of the Association of American Geographers* 65: 564–571.

Peet, R. (1988) 'Review of 'Capitalist World Development: A critique of radical development geograpshy', *Economic Geography* 64 (2): 190–192.

Santos, M. (1974) 'Geography, Marxism and underdevelopment', *Antipode* 6: 1–9.

Slater, D. (1976) 'Anglo-Saxon geography and the study of underdevelopment: Critical notes on the emergence of a new tendency', *Antipode* 8 (3): 88–93.

14 GLOBAL SHIFT (1986): PETER DICKEN

Jonathan Beaverstock

As its title suggests, the perspective of this book is global. It aims to describe and to explain the massive shifts which have been occurring in the world's manufacturing industry and to examine the impact of such large-scale changes on countries and localities across the globe. ... Ultimately, the main thread which binds the various parts of the book together is that of the effects of global industrial change. (Dicken, 1986: i)

Introduction

Peter Dicken, Emeritus Professor at the School of Environment and Development, the University of Manchester, has been one of the major 'movers and shakers' in debates in both economic geography and globalization. He has been at Manchester for over four decades, was awarded his Personal Chair in 1988 and has held distinguished positions at universities in Australia, Canada, Hong Kong, Mexico, Singapore and the United States. In 1999 he was invited to become a Fellow of the Swedish Collegium for Advanced Study in the Social Sciences and was awarded the Royal Geographical Society-Institute of British Geographer's Victoria Medal in 2001, followed by an Honorary Doctorate in Philosophy from the University of Uppsala, Sweden in 2002. Ongoing editorial positions on international journals such as *Competition and Change, Journal of Economic Geography, Global Networks, Progress in Human Geography* and *Review of International Political Economy*

provide a continuing mark of his esteem in the discipline (see Beaverstock, 2004).

Peter Dicken experienced a neo-classical economic geography upbringing in the late 1960s, spawning such 'classics' as *Location in Space: A Theoretical Approach to Economic Geography* (authored with Peter Lloyd in 1972). This book became one of the most significant texts of the period using classical and neo-classical theorists such as Christaller, Isard and Losch to explain the role of locational analysis in explaining the spatial organization of regional economic development. It was during the early 1970s, however, that Dicken began to question the orthodoxy of neo-classical locational theory and the deductive approaches that followed in the wake of the 'quantitative revolution' in economic geography. Influenced by the behavioural backlash to the 'quantitative revolution' in economic and human geography, Dicken became seduced by management research in behavioural science and organizational studies (now termed, broadly, 'Strategy'). Dicken's inductive research framework paved the way for his pioneering approach to studying the role of business decision-making processes in shaping the global distribution of economic activity (Yeung and Peck, 2003). The study of multinational enterprise (and corporations [MNCs]) very quickly became the central tenet of Dicken's work during the 1970s and early 1980s in a context of rapid MNC international restructuring and a concomitant deindustrialization process which wrought havoc with Western manufacturing industries, not least in the United

Kingdom's industrial heartlands in the North and Midlands. Dicken's highly influential writings of these times (e.g. Dicken, 1971, 1976, 1977, 1980; Lloyd and Dicken, 1972/76), quickly became the forerunners to arguably his most significant contribution to economic geography: explaining international economic change through the global behaviour and strategy of transnational corporations (TNCs) in a rapidly changing world. In essence, this research became the foundation for Dicken's seminal work, *Global Shift: Industrial Change in a Turbulent World* (1986).

Perhaps the most interesting analysis of the genesis of *Global Shift* comes from the author himself. He noted that the book took about two years to complete after being started in about 1984, and, 'at that time, "globalization", as either a focus for geographical research or as a popular topic, barely existed' (Dicken, 2004a: 513). Dicken was convinced that in order to understand industrial restructuring and territorial development at the regional scale, one had to seek explanations from what was going on at the world scale, particularly through an evaluation of the organizational strategies of TNCs. This was very true for Dicken's own 'local' work on textiles and engineering in the North West of England, an area being systematically deindustrialized and influenced significantly by world events, for example, competition from the Newly Industrializing Countries (NICs), the rise of new 'enabling' technologies and changing MNC/TNC behaviour as firms sought low-cost locations for production. Not surprisingly then, *Global Shift* was about manufacturing industries – hence the subtitle *Industrial Change in a Turbulent World*. As Dicken (2004a: 514) himself noted, '*Global Shift* was essentially a book about the global transformation of manufacturing industries and its effects on employment'. The most intriguing aspect of Dicken's thinking in the writing of *Global Shift* was his decision to not use the word 'Geography' in the title or to make any claim to be a geographer on his part. As he subsequently lamented:

> … [a]lthough the book was fundamentally geographical, I took the decision at the outset not to use the word 'geography' in the title nor even to divulge my identity as a geographer. In some ways I now feel a little ashamed of having done that. On the other hand, at that time – and, to some extent, this is still the case – most people would not have taken such a book written by a 'mere geographer' very seriously. So it proved. Sad though it may seem, I have no doubt that part of the reason the book came to be used and accepted across a wide range of social science disciplines, as a research and teaching book, was that it was approached without disciplinary preconceptions. (Dicken, 2004a: 514)

Dicken's subsequent work has elaborated many of the themes in *Global Shift*, where the 'big picture' of worldwide economic restructuring is often used as a gateway to explain industrial change, and regional and territorial (re)development. Three major examples come to mind. First, *Global Shift*'s corporation-based approach to economic restructuring has been used by Dicken as a forerunner for a significant corpus of conceptual and empirical work on 'webs on enterprise' or business networks within and between transnational corporations (see, for example, Dicken and Thrift, 1992). Second, *Global Shift* was the catalyst for research on Japan in terms of the organizational strategies of the *soga shosha* (trading companies) and patterns of foreign direct investment in Western Europe and the USA (see for example, Dicken and Miyamachi, 1998). Third, Dicken's latter-day research theorizing the significance of Global Production Networks in the world has its roots very much in the *Global Shift* discourse (see, for example, Coe *et al.*, 2004). In fact, the influence of *Global Shift* is omnipresent in many of

Dicken's writings on globalization and territorial re(development) (see, for example, Dicken, 2004b).

The book and its arguments

The beauty of *Global Shift*, which helped establish it as a key text in human geography, is its clarity of argument and rigorous interrogation of rich, in-depth empirical sources and examples taken from the economic geographies of everyday corporate life pre-the mid 1980s. Dicken argued that the rapid transformations of industrial activity since the end of the Second World War was facilitated by three major forces:

- the growth, internationalization and organizational strategies of TNCs worldwide;
- the role of nation-states and national governments through trade, investment, regional development and macroeconomic policy;
- the revolutionary impact of enabling technologies in transport, communication, production, organization and internationalization.

Combined, these three 'eclectic' explanatory forces were stitched together by Dicken to present a series of very persuasive perspectives that outlined why we should now think in terms of an interconnected and interdependent *global* economy rather than a merely inter-national economy (Dicken, 1986). Part One – 'Patterns and Processes of Global Industrial Change' – is not just about economic change as orchestrated by the geoeconomic power of TNCs, but also the role of the nation-state through instruments of macroeconomic policy, FDI incentives, trade policies (for example, GATT), the power of regional blocs (for example, the EEC), Japan, and importantly, the NICs (Newly Industrialized Countries).

In Part Two of the text, 'The Picture of Different Industries', Dicken brings forward the theoretical writings in Part One to explain international restructuring and territorial development in specific case study sectors: textile and clothing; iron and steel; motor vehicles; and electronics. Each chapter carefully explains the major processes and patterns of industrial restructuring and employment change in both a detailed and non-technical fashion. The book is completed with a series of chapters discussing the 'Stresses and strains of global industrial change' by reviewing the costs and benefits of TNCs on host countries and the effects of global economic change on the Western economics, the NICs and the Third World.

Initial reactions to the publication of *Global Shift* during the late 1980s were, on balance, very welcoming. Book reviews championed its international perspective, student-user friendly approach (particularly for the novice economic geographer) and uncomplicated, straight-talking conceptual explanations, drawn primarily from writers such as John Dunning, Stephen Hymer and Raymond Vernon (see for example reviews by Jones, 1986 and Wise, 1987). Krumme (1987: 132) lauded its eclectic approach in introducing, 'contemporary international dimensions to modern economic geography', but was slightly concerned that Dicken had underplayed the significance of international trade and had not used the opportunity in the case study chapters to focus on natural resource industries such as oil, tin, copper and global agricultural MNCs. Sharp (1987: 649) commented that the book was 'ambitious' and drew the reader's attention towards shortfalls in Dicken's appraisal of the role of technological change in driving restructuring within MNCs. But perhaps the most stinging critique came from Richard Peet's (1988) review *in Progress in Human Geography*. Peet suggested that the book's 'deliberate' eclectic approach was its major downfall because it had no significant underlying structural theory to explain its eclectic elements and their inter-relationships: '[S]o this useful book, on a topic of vital significance in the first and third worlds, lacks a theory of social structure and reaches no general conclusions' (Peet, 1988: 152).

The significance of *Global Shift*

> Peter can be accredited with putting globalization on the agenda in economic geography. (Yeung and Peck, 2003: 2)

> *Global Shift* ... is one of human geography's minuscule number of ambassadorial texts. (Olds, 2004: 510)

In 2004, the journal *Progress in Human Geography* earmarked *Global Shift* as one of its 'Classics in human geography revisited' with critical evaluations by Kris Olds (2004) and Ray Hudson (2004) (with a response from Peter Dicken himself, as noted above). Olds (2004) recognized that the very quick success of *Global Shift* was due to three main factors: timing; content; and style. As for timing, Olds noted that *Global Shift* was written at a time when international economic restructuring on a world scale was having significant impacts in local and regional economies, particularly in the North where debates about the New International Division of Labour (NIDL) (Frobel *et al.*, 1980), branch plant economies and foreign direct investment (FDI) and deindustrailization were very much in vogue. What *Global Shift* achieved in its content was a weaving together, possibly for the first time, of many of the major debates about the NIDL, TNCs, FDI, etc. in a context of international restructuring on a world scale. In terms of style, I would support Olds (2004: 509) claim, 'that the level of abstraction that Dicken employs plays a critical role in the book's attractiveness to professors (and especially students)'. *Global*

Shift's well-balanced and informative approach between theory (of the internationalization of TNCs) and empirics (a case study approach), both allied to the role of the nation-state, has produced a textbook which is highly-relevant and provides 'joined-up thinking' when explaining local and regional economic change born from a world scale perspective.

For Hudson (2004) the attractiveness and significance of *Global Shift* was evidenced in five key characteristics. First, it focused on the major actors that facilitated global economic change; nation-states, TNCs and enabling technology. Second, it displayed an overtly 'geographical' approach to explaining economic change on a world scale, teasing out the subtleties of uneven development. Third, it explained how firms' organizational forms and strategies (in different industrial sectors of the economy) generated multifaceted geographies of production, which, when combined, illustrated very clearly that processes of economic globalization were complex, uneven and highly interrelated. Fourth, it highlighted the 'footloose' tensions between TNCs and the nation-state as firm strategy continued to seek low cost locations for production. Fifth, it was an extremely informative textbook, drawing upon a wide range of ideas and data in tabular, figure and map form.

In the edited text *Remaking the Global Economy,* Jamie Peck and Henry Wai-chung Yeung (2003) pay homage to Peter Dicken and particularly the first edition of *Global Shift*. They argue that 'Dicken's most significant intervention of the 1980s was the publication of his first single-authored book *Global Shift* ... [which] ... can claim to be one of the pioneering globalization texts' (Yeung and Peck, 2004: 9). Both of these authors note that the main contributions of *Global Shift* emanated from three major debates and issues at the time of writing the book: firstly, the NIDL and associated world industrial transformation; secondly, the globalization discourse; and thirdly international business, drawn from management science. In essence, Dicken brought a critical *geographical* eye to the writing on the NIDL, following somewhat in the wake of Massey's (1984) *Spatial Division of Labour* (see Phelps, Chapter 10 this volume). *Global Shift* brought much clarity and rich empirical firm- and sectoral-specific studies to illustrate how and why the corporate strategies of TNCs reproduced the geographies of the NIDL. Moreover, Dicken was able to combine the Fordism and post-Fordism discourses with a subtle explanation of the organizational strategies of TNCs to identify the major process and patterns of global production on a world scale in the 1980s.

Yeung and Peck (2003) accordingly recognize that Dicken's *Global Shift* has made a significant contribution to the globalization discourse/debate. Dicken was at the forefront of putting the 'global' into the agenda of economic geography by studying TNCs, their international strategies, and their effects on employment and territorial development. *Global Shift* transcended the local and regional scales in both concept and empirical practice and thereby was at the forefront of implicitly advocating the global-local discourse through its passionate process-led explanations of uneven global economic change brought about by such actors as TNCs. The immense value of *Global Shift* as a globalization text is evident by the fact that '… [i]t is also one of the few geographical studies cited in complementary works on globalization by other prominent social scientists [such as] Hirst and Thompson, 1996; Held *et al.*, 1999' (Yeung and Peck, 2003: 13). *Global Shift* followed in the footsteps of a number of key texts on TNCs (e.g. Taylor and Thrift, 1982, 1986). But, according to Yeung and Peck, this is where the similarities between *Global Shift* and its competitors ended because *Global Shift* fully engaged with the major writers from the management school tradition of international business and MNC/TNC

strategy (e.g. Dunning, 1988). Dicken's skill was his ability to explore the growth, internationalization and differentiated production systems of TNCs through a geographer's distinctive gaze through the lens of international business theory.

Conclusion

The longevity of *Global Shift* (1986) is registered in the fact it has already gone to three other editions: *Global Shift: The Internationalization of Economic Activity* (1992); *Global Shift: Transforming the World Economy* (1998) and *Global Shift: Reshaping the Global Economic Map* (2003). Whilst each edition has added significant value from the previous one in terms of building theoretical understanding and empirical quality, the spine of the meta-narrative remains just as innovative in the 2003 edition as it did in 1986. The 1992 edition brought services into the equation with a dedicated chapter focused on their internationalization and significance in making the world economy. In addition, in Dicken's appraisal of the processes of economic change, more emphasis was given to the network of relationships that exist within and between TNCs. By the 1998 and 2003 editions, Dicken had refined and embellished more the processes of global shift and in doing so produced one of the classic explanations accounting for the growth, shape, internationalization and network organizational form of TNCs in the contemporary world economy. The 2003 edition saw the addition of a new chapter on the distribution industries; in January 2007, Sage published the 5th Edition of the text: *Global Shift: Mapping the Changing Contours of the World Economy*, with companion resources available at the website www.sagepub.co.uk/dicken.

Notwithstanding the initial critical maulings from Peet (1988) and others, *Global Shift* has stood the test of time and been instrumental in the development of many key strands of economic geography since its initial publication.

Four, of many, come to mind. First, *Global Shift* shaped the way geographers and other social scientists embraced the globalization discourse. One cannot underestimate the role the book played in moving this debate/process forward into the 1990s. Second, *Global Shift* very clearly illustrated the dynamic and changing role of NICs in the world economy, particularly Asian tiger economies, including Japan. Third, *Global Shift* lauded the 'actor' approach (the TNC) in understanding economic change in the world economy. Dicken's close analysis of international business systems, combined with a spatial perspective, has provided the forerunner for contemporary work on such topics as: production (and commodity) chains, organizational networks and firm–buyer relationships. Fourth, linked to the third, *Global Shift* has fostered recent work on Global Production Networks and relationality in economy (see for example, Coe *et al.*, 2004; Dicken *et al.*, 2001). But, more than anything else, *Global Shift* has encouraged a global sense of space and place when explaining the global-local economic nexus in understanding contemporary maps of world activity. In fact, Nigel Thrift's strap-line for the recent edition sums up *Global Shift's* influence on geography, 'Global Shift just keeps on getting better. There is no other source that gives you the full story of globalization in such a fluent and authoritative way. Not just recommended but essential'.

Secondary sources and references

Beaverstock, J.V. (2004) 'Peter Dicken', in P. Hubbard, R. Kitchin and G. Valentine (eds), *Key Thinkers on Space and Place.* London: Sage, pp. 108–112.

Coe, N., Hess, M., Yeung, H.W.C., Dicken, P. and Henderson, J. (2004) 'Globalizing regional development: A global production networks perspective', *Transactions of the Institute of British Geographers* 29 (4): 468–484.

Dicken, P. (1971) 'Some aspects of the decision-making behaviour of business organizations', *Economic Geography* 47 (3): 426–437.

Dicken, P. (1976) 'The multiplant business enterprise and geographical space', *Regional Studies* 10 (4): 401–412.

Dicken, P. (1977) 'A note on locational theory and the large business enterprise', *Area* 9 (2): 139–143.

Dicken, P. (1980) 'Foreign direct investment in European manufacturing industry', *Geoforum* 11 (2): 289–313.

Dicken, P. (1986) *Global Shift: Industrial Change in a Turbulent World.* London: Harper & Row.

Dicken, P. (1992) *Global Shift: The Internationalization of Economic Activity.* London: Paul Chapman Publishing.

Dicken, P. (1998) *Global Shift: Transforming the World Economy.* London: Paul Chapman Publishing.

Dicken, P. (2003) *Global Shift: Reshaping the Global Economic Map in the 21st Century.* London: Sage.

Dicken, P. (2004a) 'Author's response', *Progress in Human Geography* 28 (4): 513–515.

Dicken, P. (2004b) 'Geographers and globalization: (yet) another missed boat', *Transactions of the Institute of British Geographers* 29 (1): 5–26.

Dicken, P. and Miyamachi, Y. (1998) 'From noodles to satellites: The changing geography of Japanese sogo shosha', *Transactions of the Institute of British Geographers* 23 (1): 55–78.

Dicken, P. and Thrift, N. (1992) 'The organization of production and the production of organization: why business enterprise matter in the study of geographical internationalisation', *Transactions of the Institute of British Geographers* 17 (1): 101–128.

Dicken, P., Kelly, P.F., Olds, K. and Yeung, H.W. (2001) 'Chains and networks, territories and scales: towards a relational framework for analysing the global economy', *Global Networks* 1 (1): 99–123.

Dunning, J.H. (1988) *Explaining International Production.* London: Unwin Hyman.

Frobel, F., Heinrichs, J. and Kreye, O. (1980) *The New International Division of Labour.* Cambridge: Cambridge University Press.

Hudson, R. (2004) 'Commentary 2', *Progress in Human Geography* 28 (4): 511–513.

Jones, P. (1986) 'Book review: Global Shift', *Geography* 71 (4): 377.

Krumme, G. (1987) 'Book review: Global Shift', *Environment and Planning A* 19 (1): 132–133.

Lloyd, P.E. and Dicken, P. (1972 and 1976) *Location in Space* (1st and 2nd editions). New York: Harper & Row.

Massey, D. (1984) *The Spatial Division of Labour.* Basingstoke: Macmillan.

Olds, K. (2004) 'Commentary 1', *Progress in Human Geography* 28 (4): 507–511.

Peck, J. and Yeung, H.W.C. (eds) (2003) *Remaking the Global Economy.* London: Sage.

Peet, R. (1988) 'Book review: Global Shift', *Progress in Human Geography* 12 (1): 151–152.

Sharp, M. (1987) 'Book review: Global Shift', *International Affairs* 63 (4): 649.

Taylor, M. and Thrift, N. (eds) (1982) *The Geography of Multinationals.* London: Croom Helm.

Taylor, M. and Thrift, N. (eds) (1986) *Multinationals and the Restructuring of the World Economy.* London: Croom Helm.

Wise, M.J. (1987) 'Book review: Global Shift', *The Geographical Journal* 153 (2): 274–275.

Yeung, H.W.C. and Peck, J. (2003) 'Making global connections: A geographer's perspective', in J. Peck and H. Wai-chung Yeung (eds) (2003) *Remaking the Global Economy.* London: Sage, pp. 3–23.

15 THE CONDITION OF POSTMODERNITY (1989): DAVID HARVEY

Keith Woodward and John Paul Jones III

The experience of time and space has changed, the confidence in the association between scientific and moral judgements has collapsed, aesthetics has triumphed over ethics as a prime focus of social and intellectual concern, images dominate narratives, ephemerality and fragmentation take precedence over eternal truths and unified politics, and explanations have shifted from the realm of material and political-economic groundings towards a consideration of autonomous cultural and political practices. (Harvey, 1989: 328)

Introduction

David Harvey's (1989) *The Condition of Postmodernity: An Enquiry into the Origins of Cultural Change* is more than a key text in geography: its popularity and significance is unmatched *outside* of the discipline. A few minutes with Google™ Scholar will affirm that no academic book discussed in this volume has been as widely cited as *Condition of Postmodernity*. For less systematic evidence, students need only visit the offices of professors who work in other fields; they'll often find that *Condition of Postmodernity* is the only book by a geographer on their shelves. The forcefulness of Harvey's argument led Terry Eagleton, the renowned Marxist literary theorist, to offer the following assessment, printed on the back cover of *Condition*:

Devastating. The most brilliant study of postmodernity to date. David Harvey cuts beneath the theoretical debates about postmodernist culture to reveal the social and economic basis of this apparently free-floating phenomenon. After reading this book, those who fashionably scorn the idea of a 'total' critique had better think again. (Eagleton, 1989: np)

But the book itself is only part of the story of its popularity. Another is that strange conjuncture of intellectual thought, cultural trends, economic transformations, and political developments that in the 1980s came to be known as 'postmodernism.' It is hard for those whose intellectual awakening came in the late 1990s or later to have a sense of that era – of the immediacy of opportunities and dangers it seemed to present – but consider this: for several hundred years something that came to be called 'modernity' developed apace. And then, like tracking the changing temperature of time itself, there emerged a widespread feeling that modernity's cherished moorings – a faith in human rationality and logical communication, in economic, political, and social progress, in science, technology, and aesthetic coherence, and in just and ethical systems of valuation and judgment – were being unhinged to such an extent that the world, especially the West, was entering a new era. Though Harvey draws on his always-keen geographical imagination in analyzing postmodernism, his account of these shifts goes to issues much larger than the discipline of geography, and this helps explain why his book has been so widely read. *Condition of Postmodernity* touched a chord to which academics of many stripes were attuned.

This essay on *Condition of Postmodernity* covers its argument, impact, and critical reception. But before we move forward, there are three preliminaries to address. First, *Condition of Postmodernity* is unlike the books by Harvey that immediately preceded (Harvey, 1982 – see Castree, Chapter 8 this volume) and followed (Harvey, 1996) it, both of which he nearly abandoned in frustration, in that it nearly 'wrote itself.' Harvey reports that the writing came so easily that it 'poured out lickety-split' (Harvey, 2002: 180). Harvey's previous work on the urbanization of capital, on the history of Second Empire Paris and modern Baltimore, and on space and time within dialectical materialism were foundational for *Condition of Postmodernity*. The book amplifies an analysis in a 1987 essay he published in the radical geography journal, *Antipode*, in which he argued, in line with an earlier essay by Fredric Jameson (1984), that 'post-modernity is nothing more than the cultural clothing of flexible accumulation' (Harvey, 1987: 279). That piece ended with a challenge befitting Harvey's intellectual debt to Marxism: 'A critical appraisal … of the cultural practices of postmodernity … appears as one small but necessary preparatory step towards the reconstitution of a movement of global opposition to a plainly sick and troubled capitalist hegemony' (Harvey, 1987: 283). It was Harvey's friend and long-time editor, John Davey, who persuaded him to make that appraisal.

A second thing to know is that *Condition of Postmodernity* – along with Jameson's denser but similarly minded *Postmodernism, or, the Cultural Logic of Late Capitalism* (1991) – greatly contributed to making its object of analysis passé. This is not to say that there were not already strong critics of postmodernism, and perhaps even a sense of intellectual and cultural exhaustion, by 1989. As Harvey (1989: ix) noted in his Preface: 'When even the developers tell an architect like Moshe Safdie that they are tired of it, then can [The philosophers] be far behind?' Harvey's own strategy was to *historicize, locate,* and *explain* postmodernism, and there are few things more disabling for a movement that fancies detached moorings. After *Condition,* as Eagleton noted, the *foundations* of an apparently free-floating phenomenon were established. Wind out of sails, the ship was grounded; postmodernism's themes live on, but under different banners.

Third, in our opinion *Condition of Postmodernity* should be read in conjunction with the text that followed it, *Justice, Nature and the Geography of Difference* (1996). A much misunderstood book, *Justice* complements its predecessor: by specifying a suite of ontological questions that lay dormant in *Condition of Postmodernity*; by laying out in clear detail a dialectical analytic that underwrites Harvey's approach to explanation; and by responding to critics of *Condition of Postmodernity* by addressing its widely acknowledged sublimation of gendered and raced social relations. Harvey's *Justice* also presages the current interest by geographers and others in ethics and responsibility, and on how to theorize the relationship between culture (social life) and the natural environment. Whereas the book we describe here is largely critique, *Justice* helps readers understand more fully how that critique is grounded, while responding to lapses in some of *Condition of Postmodernity's* arguments.

In writing about postmodernism, Harvey once affirmed Pierre Bourdieu's injunction that, 'Every established order tends to produce the naturalization of its own arbitrariness' (quoted in Harvey, 1987: 279). For all its uncertainty, multiplicity, and disorderliness, the age of postmodernism was very nearly one of an 'established order.' It was Harvey's mission to unmask that naturalization. In the process, *Condition of Postmodernity* became part of the time–space conjunction it analyzed, further naturalizing the book as, in Eagleton's terms, devastating and brilliant.

The argument

For many reasons, it is important to read *Condition of Postmodernity* as a text devoted principally to the critique of a system rather than the promotion of a coherent alternative. While it is undeniable that Harvey never strays from the project of spatializing the political economy, a move that spans almost the entirety of his oeuvre, here the positive, revolutionary, and utopian aspects of Marxism are sent to the background, as something that sits behind, but nevertheless frames, the topic at hand. True to its title, the book provides a critical analysis of economic and cultural *conditions* specific to and definitive of the last quarter of the twentieth century. Importantly, Harvey recognizes that a double meaning hides within the idea of 'condition': it signals the state of actual, existing *things* surrounding us and making up the world but, at the same time, it also indicates the historical tendencies driving global *processes*. Put another way, 'condition' implies both the *condition of things* and *that which conditions things*. It is within this double formulation that surfaces the beginnings of the ontological development that will come to maturity in *Justice, Nature and the Geography of Difference*: a condition is, at once, a state of Being *and* a process of Becoming.

Historically, the 'condition of postmodernity' is said to have developed out of (or to have been a break with) the vast collection of Western philosophical, artistic, and scientific theories that developed during the period known as 'modernism.' Though beginning with the Enlightenment, this historical era gained ground through the establishment of scientific positivism; the growth, spread, and techno-practical coherences of industrial capitalism; and the development of the democratic state form. These were in no way discrete historical events or processes; rather, each informed the other. Moreover, they helped map out a human-centered world aimed at the development of free and autonomous human agents: rational economic citizens naturally embracing science, capitalism, and democracy. By the mid-twentieth century, however, the ideals of modernity had been pushed into crisis by the increasingly glaring inequalities that accompanied the development of capitalism and by the ever-greater alienation fostered by the violence and devastation of the two World Wars. Artists, philosophers, and even scientists increasingly turned to fragmented, alienated, and relativistic representations of the world, revealing a growing dissatisfaction with appeals to the foundationalism that had been the cornerstone of modernist thinking.

Postmodernity met this discontent with several accounts of difference, positionality, and situatedness that appeared to ring the death-knell for aging visions of a world rooted in essentialism, totalization, and universality. One of the key moments in this transformation was Jean-François Lyotard's *The Postmodern Condition* (1984) – from which Harvey's book derived its name and to which it serves as a response. Here, Lyotard called for a rejection of the 'grand narratives' of modernity, two of which were especially suspect in the postmodern critique: the assumed total autonomy of the individual (liberalist humanism, free market entrepreneurialism), and the linear deterministic progression of history (Marxist socialism, scientific progress). Thinkers such as Michel Foucault, Jacques Derrida, and Lyotard argued that such notions did not reflect any necessity within reality or the 'nature' of things so much as the influence of power and discourse in the ways we know and understand the world (Dixon and Jones, 2004). Likewise, language, politics, and even identities became matters not of universals, but of particularity, contingency, and difference.

Given this apparent break with the foundations of modernity, what possibilities remain for the collective politics that Harvey and other Marxists find necessary for undermining capitalism? His solution was to hold his ground, reanalyzing the relation between modernity and postmodernity. He concludes that the latter, in fact, does not represent a break with the former, but rather its continuation, with changes marking adjustments to transformations in capitalist production and consumption. For Harvey, modernity is inseparable from the processes and institutions devoted to the accumulation of capital and the utilization of labor, reaching its point of inflection with the advent in 1914 of Fordism. Initiated by Henry Ford's introduction of the 'five dollar, eight hour day as recompense for workers manning the automated car-assembly line he had established' (Harvey, 1989: 125), Fordism was sealed in the post-World War II era as a social compact among capitalists, labor unions, and the social welfare state. The macroeconomics of Fordism was globalized under the Bretton Woods agreement of 1944, which 'turned the dollar into the world's reserve currency and tied the world's economic development firmly into US fiscal and monetary policy' (Harvey, 1989: 136). Accompanying this agreement was the opening of global markets to American corporate interests, and eventually Fordism began to spread throughout the globe.

By the mid-1960s, however, a number of national and regional markets had arisen to challenge 'United States hegemony within Fordism to the point where the Breton Woods agreement cracked and the dollar was devalued' (Harvey, 1989: 141). Drawing from his earlier theories established in *Limits to Capital* (1982; see Castree, Chapter 8 this volume), Harvey points to the unraveling of

Fordism in the 1960s and 1970s: a system too rigid and contradictory to put off crises of over-accumulation, it was inexorably being supplanted by a new, post-Fordist or flexible form of accumulation. Flexibility was sectoral insofar as capital was moved to invest in service industries; it was technical in the shift toward more fluid labor agreements and outsourcing arrangements; and it was geographical in capitalism's ever demanding need to 'spatially fix' its crises by mobilizing in ways that lower costs, open new markets, and increase profits. Flexibility emphasized greater adventurism on the part of the capitalist through the production of mobile, short-lived commodities while, for the worker, whose own labor is sold *as* a commodity, this meant new forms of exploitation as promises of future employment were increasingly broken, which in turn fostered increased transience and 'nomadism' within the laboring class.

Critically, Harvey argues that, as the distances and times it took to accumulate capital and circulate commodities shrunk, our *experience* of space and time similarly compressed. What is more, postmodernity's rise at this juncture – as an intellectual, architectural, artistic, and cultural movement – was not coincidental, for the sea-change called postmodernism is the direct result of these experiential dislocations. So, while previous representations of postmodernity might have argued that the moment was fundamentally the product of *cultural* transformations (from which economic changes, like the rise of entertainment industries or the growth of gentrification, then followed), Harvey's analysis of the post-Fordist political economy turned this formulation upon its head. Making culture the shadow of economic processes, he explained in no uncertain terms that: 'the emphasis upon ephemerality, collage, fragmentation, and dispersal in philosophical and social thought *mimics* the conditions of flexible accumulation' (Harvey, 1989: 302, our emphasis). Harvey illustrates this causal reversal by examining key components of Western culture, drawing upon: (a) the recent history of the American city-scape, where he assesses several exemplary postmodern urban designs, including the spectacle-producing Disneyfication of Baltimore Harbor; and (b) the loss of depth, meaning, and history in art and aesthetics, echoing a widespread emphasis on 'the values and virtues of instantaneity … and of disposability' (Harvey, 1989: 286) as capitalism moves from Fordist mass production to flexibility.

Perhaps the most jarring aspect of Harvey's argument is the suggestion that a great variety of developments in recent progressive politics – such as feminism, anti-racism, and queer activism – by virtue of their emphases upon an apparently relativistic politics of positionality, proceed in the spirit of these recent processes of capitalist development. Envisioning a postmodern identitarian politics that shares more commonalities with post-Fordist capitalism than with Marxist anti-capitalism (Harvey, 1989: 65), he suggests that postmodern strategies and argumentation drawn upon by identity politics after the cultural turn may be only *apparently* progressive:

> … postmodernism, with its emphasis upon the ephemerality of *jouissance*, its insistence upon the impenetrability of the other, its concentration upon the text rather than the work, its penchant for deconstruction bordering on nihilism, its preference for aesthetics over ethics, takes matters too far. It takes them beyond the point where any coherent politics are left, while that wing of it that seeks a shameless accommodation with the market puts it firmly in the tracks of an entrepreneurial culture that is the hallmark of reactionary neoconservativism. (Harvey, 1989: 116)

Thus, while politics of positionality may seem progressive, Harvey asserts that such fragmented strategies are in fact openings for, if not inspired by, the equally fragmented practices of accumulation and production in contemporary capitalism and, importantly, their attendant transformations of the spaces that we daily encounter. In so arguing, Harvey makes the spaces and processes of post-Fordist capitalism the *conditions* for culturally inflected politics: 'Aesthetic and cultural practices are particularly susceptible to the changing experience of space and time precisely because they entail the construction of spatial representations and artifacts out of the flow of human experience. They always broker between Being and Becoming' (Harvey, 1989: 327).

The impact of the *Condition of Postmodernity*

As mentioned at the outset of this chapter, *Condition of Postmodernity* was widely read in many disciplines, and it has had a lasting impact. In geography, the book's most lasting mark ensued from Harvey's efforts to connect economic and cultural analysis. For many reasons that well predate the publication of *Condition of Postmodernity*, economic and cultural geography had largely developed independently of one another since the postwar period. The former grew in sophistication under the dual influence of both spatial scientific and Marxist theories (see Barnes, 1996; see Kelly, Chapter 23 this volume) while the latter was either practiced as nave empiricism (e.g., cultural geographies of housetypes and the like) or developed inspiration from humanistic geography (e.g., Tuan, 1977; see Cresswell, Chapter 7 this volume). Now, it is true that there is a history of linking economic and cultural phenomena in critical theory: one need only point to the base-superstructure model in Marxism, or to the efforts of thinkers such as E. P. Thompson, Raymond Williams, Theodore Adorno, and Stuart Hall, among others. But in geography at the time of *Condition's* publication, there were only a handful of geographers who were attempting, as Harvey was, to bring together the traditions of cultural interpretation and political economic analysis. Among the notables in this period was Denis Cosgrove, whose *Social Formation and Symbolic Landscape* (1984 – see Gilbert, Chapter 12 this volume) brilliantly wove together political, economic, and cultural analysis. And Harvey too was an early weaver of political economy and cultural interpretation in geography: one need only look at his now classic analysis of the political symbolism of the Basilica of Sacré-Coeur in Paris (Harvey, 1979).

In any event, the more general point is that however much Harvey would be later criticized for what were seen as one-sided readings of cultural texts (see below), and, for that matter, however much one might disagree with his assessment that economic change drives cultural responses, *Condition of Postmodernity* stands as a model effort to bring together two subfields largely separated due to their different objects of analysis (paintings versus factories) and the theories and methodologies typically brought to bear on them. In terms of analytic strategies, then, part of *Condition's* value comes from illustrating the connections between economy and culture.

Since its publication, economic and cultural geography have become much more closely aligned if not integrated, and while it is impossible to thread causality back to the book's appearance, *Condition of Postmodernity* must nonetheless be given its due. Today, many economic geographers actively embrace the subfield's 'cultural turn' by investigating – usually in dialectical fashion – the intersection between cultural forms and political-economic processes, often analyzing texts and visual media with the tools of content and discourse analysis favored by cultural geographers. Although quite different in their approach to the causal relations involved, we find this sort of analysis exemplified in the work of: Linda McDowell, who integrates cultural analysis, economic geography, and feminism in *Capital Culture: Gender at Work in the City* (1997); Nigel Thrift, who hones in on capitalism's own 'cultural circuit' in *Knowing Capitalism* (2005); and Allen Scott, whose *The Cultural Economy of Cities* (2000) examines the economic bases of the culture industries in Los Angeles and other world cities. This is not to say that *Condition* is responsible for the growth of representational or discursive analyses of the economy during the last decade of the twentieth century, for Harvey's own stance with respect to 'cultural political economy', as it is sometimes called, is clear enough. It is, rather, to point to the fact that however contentious economic geography's cultural turn has been, *Condition*, by virtue *of* its integration of economy and culture, opened up new territories in geography. Whether Harvey would approve of the circuitous routes taken to get to the intersection is, however, a different matter.

Critical reactions

Of course, *Condition of Postmodernity* was not uniformly welcomed in geography, nor for that matter was it praised in all quarters of social and cultural theory. This is to be expected: geography, as this volume shows, is a highly differentiated and often contentious terrain, and many of the lines of conflict within the field are refractions of similar debates within theory more generally. At worst, there were geographers and others who, in claiming Harvey to *be* a postmodernist, demonstrated that they never read the book, much less its back cover. At its best, at the time of *Condition of Postmodernity's* appearance, two of the most important intellectual fault lines in critical geography (and elsewhere) were between what was perceived to be an overly structuralist, totalizing, and economistic Marxism (Duncan and Ley, 1982), on the one hand, and poststructuralism and feminism, on the other hand. These last two – which are related to but not subsumed by the more substantive tensions between economic and cultural geography, as discussed above – were sometimes overlapping critical injunctions that together provided a tense intellectual field for the reception of the book, and nowhere is this better illustrated than in two lengthy essays, one by Rosalyn Deutsche (1991) and one by Doreen Massey (1991), both of which drew on poststructuralist feminism to offer withering critiques.

Deutsche began her critique by accusing Harvey of relying on a masculinist and ocularcentric epistemology that unreflexively professes confidence in the ability to clearly grasp causal connections free of any complications that might be introduced by the viewer's social positionality. This 'totalizing' view, she maintained, underlies Harvey's deployment of a rigid Marxist analytic aimed at taming an unruly postmodernism filled with difference and possibility. It also explains his failure to recognize any limits in his perspective, as well as his lack of acknowledgement of both feminist work on postmodernism (not in any way an easy combination: see Nicholson, 1990) and feminist representational theory, particularly as it circulated within the domain of art. As for the latter, she offered a stinging criticism of Harvey's reading of the photographic self-portraits of artist Cindy Sherman: where Harvey saw in Sherman's many disguises evidence for a depthless postmodern fetishism, Deutsche read them as critical commentaries on modernist artistic theories and their emphases on the individual artist and 'his' invocation of universal truths; for Deutsche, Sherman's portraits disrupted such authorial notions by pointing to the social construction of meaning in artistic representations, while simultaneously questioning the 'reality effect' of documentary photography. On this point

Deutsche's poststructuralism departs from Harvey's geographic materialism, tapping into what had become a seemingly endless point-counterpoint between materiality and representation in the study of socio-spatial life (Dixon and Jones, 2004). As Deutsche had it:

> Reality and representation mutually imply each other. This does not mean, as it is frequently held, that no reality exists or that it is unknowable, but only that no founding presence, no objective source, or privileged ground of meaning ensures a truth lurking behind representations and independent of subjects. Nor is the stress on representation a desertion of the field of politics ... any claim to know directly a truth outside representation emerges as an authoritarian form of representation employed in the battles to name reality. (Deutsche, 1991: 21)

Massey's (1991) critique echoes elements found in Deutsche's while redoubling on Harvey's limited engagement with feminist analyses of patriarchy. She begins by quoting a now famous question posed by Nancy Hartsock (1987): 'Why is it, exactly at the moment when so many of us who have been silenced [under modernism] begin to demand the right to name ourselves, to act as subjects rather than objects of history, that just then the concept of subjecthood becomes [under postmodernism] "problematic"?' (quoted in Massey, 1991: 33, brackets added). She locates the rise of the postmodern not, as Harvey does, within the coordinates of time–space compression, but in two opposing trends: progressive political activity marshaled around difference 'in fields such as feminist studies, ethnic studies, and Third World studies' (Massey, 1991: 34), and the competitive jostling for position among career minded academics. On balance, however, Massey offers a more hopeful, feminist reading of postmodernism, while at the same time affirming a commitment to retaining 'strong aspects of what characterises the modernist project, most particularly its commitment to change, hopefully progressive' (Massey, 1991: 52). Nonetheless, modernity cannot be let off the hook:

> … the experience of modernism/modernity as it is customarily recorded, the production of what are customarily assumed to be its major cultural artefacts, and even its customary definition, are all constructed on and are constructive of particular forms of gender relations and definitions of masculinity and of what it means to be a woman. This is not ('just') to say that modernism was or is patriarchal (this would hardly be news, nor differentiate it from many other periods in history); it is to say that it is not possible fully to understand modernism without taking account of this. To return more directly to Harvey, modernism is about more than a particular articulation of the power relations of time, space, and money. (Massey, 1991: 49)

The following year, Harvey (1992) responded to both Deutsche and Massey (but, significantly, writing in the radical geography journal, *Antipode*, and not in *Environment and Planning D: Society and Space*, the journal in which their critiques appeared). He began by acknowledging his regret for not integrating more feminist work into *Condition*, noting that, had he done so, the argument would have been strengthened rather than diluted. But Harvey largely stuck to his guns, employing his own differencing strategy whereby Deutsche's and Massey's analyses were particularized as emerging from one type of feminism, and not the one that suited his theory. In brief, Harvey has much more in common with the socialist feminism of Nancy Hartsock (1987) than with the poststructuralist feminism of Deutsche, and his response is at pains to point out the dangers

of what he perceives as a relativistic feminism that, while addressing difference, lacks the strong evaluative criteria to distinguish between socially and politically important axes of identification and insignificant ones: 'Lacking any sense of the commonalities which define difference, Deutsche is forced to adopt an undifferentiated, homogenizing, a-historical and in the end purely idealist notion of difference …' (Harvey, 1992: 310). The charge of idealism, moreover, runs through Harvey's dismissal of Deutsche's poststructuralism:

> Deutsche may find my view that there are social processes at work which are real unduly limiting, but I find … the view that all understanding is preconstituted, not with reference to a material world of social processes but with reference to media images and discourses about that world, not only even more limiting but downright reactionary since it leaves us helpless victims of discourse determinism. (Harvey, 1992: 316)

Harvey was even less sympathetic to Massey's essay, accusing her in an obviously angry response of invoking a 'flexible feminism' in what he saw as a personal and opportunistic attack based on the fallacious assumption (technically, a circumstantial ad hominem) that 'whatever the male gaze lights upon is bound to be given an exclusively masculine and therefore sexist reading' (Harvey, 1992: 317).

Conclusion

In this chapter we have discussed the arguments in, impact of, and responses to *Condition of Postmodernity*. Each has circulated around issues of *content*; that is, they pivot on Harvey's analysis of modernity's passage to postmodernity, flexible accumulation's role in time–space compression, the role of class relative to other aspects of social difference, and the proper theorization of economy and culture. Another way to approach the book is through its 'mode of explanation': how does Harvey's analysis *work*? To answer that question we turn in this conclusion to the topic of ontology: theories of what the world is like, how it operates, and how as a consequence we might understand and explain it.

At the outset, it is important to emphasize that discussions of ontology are never far removed from the politics of research more generally, and this is nowhere more apparent than in critical geography, which is currently engaged in a spirited debate over the status of 'leftist' thinking within the field. On the one hand, these debates are about differences in and commitments to leftist thought with respect to the theory and praxis needed to move forward, that is, to confront and potentially overturn contemporary capitalism. On the other hand, they express deep-seated differences in ontology, and in particular between the dialectical approach of critical realism and various strains of anti-essentialist poststructuralism.

Critical realists (Sayer, 1992) hold that events are caused through the operation of necessary and contingent forces. These are theorized as embedded within a depth ontology, wherein more general causal forces (e.g., capitalism) are said to work in conjunction with contingent and contextually bound ones (e.g., local political culture, the particulars of context-specific gender relations). Harvey's book offers an example of a depth ontology, as time–space compression is positioned as a mediating mechanism between the underlying structural forces of capitalism and the resulting, surface level cultural forms of intellectual trends, art and architecture, and politics. *Condition of Postmodernity,* while not explicitly referencing critical realism, demonstrates its explanatory power, and

while critical realism is not by necessity Marxist, it nevertheless complements Harvey's historical-geographic materialist analysis because of its focus on internal rather than external relations among social phenomena (Sayer, 1992).

By contrast, poststructuralists reject the notion of the structuring systems that characterize depth ontology. Either the world is 'overdetermined' (Gibson-Graham, 1996), a model in which causal forces are so mutually co-constituted that they are theoretically inseparable, or reality is itself always once removed through thought and language, and thus by the processes of categorization that name and organize the world. In both forms, poststructuralists in the 1990s came to argue on behalf of theoretical agnosticism with respect to ontology, preferring instead to prioritize epistemology, that is, how we come to know the world rather than what makes up the world. Quite simply, these authors – of whom Deutsche (1991) is a good example – maintain that epistemology 'trumps' ontology (Dixon and Jones, 2004). It should be clear that Harvey has little patience for this group of theorists.

More recently, however, numerous poststructuralists using various versions of affect theory, actor-network theory, non-representational theory, and site ontology, have reasserted the importance of materialist ontological analysis (see, for example, Geoforum, 2004). Although Harvey will be bemused to see himself described in parallel with these trends, he nevertheless shares an affinity with them by virtue of his longstanding dedication to an ontology framed by historical-geographical materialism. In this sense, to the extent that one outcome of the post-postmodern era is a vital and renewed engagement with ontology, then one of this book's lasting legacies will have been its showcasing of the limits of a postmodernism preoccupied by epistemology and discourse.

Secondary sources and references

Barnes, T. (1996) *Logics of Dislocation: Models, Metaphors and Meanings of Economic Space*. New York: Guilford.

Cosgrove, D. (1984) *Social Formation and Symbolic Landscape*. London: Croom Helm.

Deutsche, R. (1991) 'Boys town', *Environment and Planning D: Society and Space* 9: 5–30.

Dixon, D.P. and Jones III, J.P. (2004) 'Poststructuralism', in *A Companion to Cultural Geography*, James D. Duncan, Nuala Johnson, and Richard Schein (eds), pp. 79–107. Oxford: Blackwell, pp. 79–107.

Duncan, J. and Ley, D. (1982) 'Structural Marxism and human geography: a critical assessment', *Annals of the Association of American Geographers* 72: 30–59.

Geoforum (2004) 25 (6): 675–764.

Gibson-Graham, J.K. (1996) *The End of Capitalism (as we knew it): A Feminist Critique of Political Economy*. Cambridge: Blackwell.

Hartsock, N. (1987) *Money, Sex, and Power: Toward a Feminist Historical Materialism*. Evanston: Northwestern University Press.

Harvey, D. (1979) 'Monument and myth', *Annals of the Association of American Geographers* 69: 362–381.

Harvey, D. (1982) *The Limits to Capital*. Oxford: Blackwell.

Harvey, D. (1987) 'Flexible accumulation through urbanization: reflections on "postmodernism" in the American city', *Antipode* 19: 260–286.

Harvey, D. (1989) *The Condition of Postmodernity: An Enquiry into the Origins of Cultural Change*. Oxford: Blackwell.

Harvey, D. (1992) 'Postmodern morality plays', *Antipode* 24: 300–326.

Harvey, D. (1996) *Justice, Nature and the Geography of Difference*. Oxford: Blackwell.

Harvey, D. (2002) 'Memories and desire', in P. Gould and F.R. Pitts (eds), *Geographical Voices*. Syracuse: Syracuse University Press, pp. 149–188.

Jameson, F. (1984) 'Postmodernism, or, the cultural logic of late capitalism', *New Left Review* 146: 53–92.

Jameson, F. (1991) *Postmodernism, or, the Cultural Logic of Late Capitalism*. Durham: Duke University Press.

Lyotard, J.F. (1984) *The Postmodern Condition: A Report on Knowledge* (Geoff Bennington and Brian Massumi, trs). Minneapolis: University of Minnesota Press.

Massey, D. (1991) 'Flexible sexism', *Environment and Planning D: Society and Space* 9: 31–57.

McDowell, L. (1997) *Capital Culture: Gender at Work in the City*. Oxford: Blackwell.

Nicholson, L.J. (ed) (1990) *Feminism/Postmodernism*. New York: Routledge.

Sayer, A. (1992) *Method in Social Science: A Realist Approach* (2nd edition). New York: Routledge.

Scott, A.J. (2000) *The Cultural Economy of Cities*. London: Sage.

Thrift, N.J. (2005) *Knowing Capitalism*. London: Sage.

Tuan, Y.F. (1977) *Space and Place*. Minneapolis: University of Minnesota.

16 POSTMODERN GEOGRAPHIES (1989): EDWARD SOJA

Claudio Minca

For at least the past century, time and history have occupied a privileged position in the practical and theoretical consciousness of Western Marxist and critical social science. [...] Today, however, it might be space more than time that hides consequences from us, the 'making of geography' that provides the most revealing tactical and theoretical world. This is the insistent promise and promise of postmodern geographies. (Soja, 1989: 1)

Introduction

'All geographers should read this book', urged Michael Dear in his review of *Postmodern Geographies* that appeared on the *Annals of the Association of American Geographers* in 1991. And if not all, then certainly a majority of the geographers writing in those years on the relationship between geography and social theory did indeed read – and engage with – Ed Soja's book. *Postmodern Geographies*, together with David Harvey's *The Condition of Postmodernity* published that same year (see Woodward and Jones, Chapter 15 this volume), is a work that has exercised a profound influence on a whole generation of critical human geographers (in English-speaking academia but also well beyond) and whose impact on the discipline is visible even well over a decade after its release. It is enough to think of the widespread adoption in geography of the work of Foucault and Lefebvre – two key reference points in Soja's call for a critical human geography – or to the currency

gained by the concept of *socio-spatial dialectic* to realize how pervasive this influence has been. Soja was one of the first writers to 'import' into geography what he referred to as the *new French spatial school*, and his book was a disciplinary milestone in opening a dialogue with critical social theory.

The year 1989 was the last of a vibrant and, in many ways, revolutionary intellectual decade for human geography and for the 'spatial turn' in the social sciences more broadly. The publication of *Postmodern Geographies* came at a critical juncture in geography's recent history: a moment in which 'postmodern' perspectives were affirming themselves but had not yet been completely metabolized by a geographical community that was becoming – at least in the English-speaking world – increasingly 'critical'. The book's seductive title perfectly captured the spirit of the times, while the subtitle ('The Reassertion of Space in Critical Social Theory') spoke to readers outside of the discipline, helping make this work a classic not only in geography but also in what we can refer to by way of short-hand as 'postmodern studies'.

Postmodern Geographies was, in many ways, something entirely new. Michael Dear's (1991: 652) review of the book argued that 'In his reconstruction of Los Angeles [Soja] invents a new way of writing about the city', concluding that 'within geography, I would compare Soja's achievement favourably with the other pivotal texts of recent decades: Haggett's *Locational Analysis and Human Geography* (1965); Harvey's *Social Justice and the City* (1973), and Gregory's *Ideology, Science and Human Geography* (1978)'. But the book's novelty – and daring – made it also the object of critique or, in some cases, a pointed silence. *Postmodern Geographies* was, in fact, curiously ignored by many key geographical journals, especially those based in the UK, where very few reviews of the book appeared (but see Dear, 1991; Eflin, 1990; Rose, 1991; Smith, 1991; from outside the discipline see, for example, MacLaughlin, 1994; Resch, 1992). Such silence was surprising, especially for someone like myself who at the time was engaged in discussions about the book with colleagues in Italy and beyond, and enthusiastically leading successive generations of students through Soja's work.

The reason why I mention the somewhat intricate academic trajectories of *Postmodern Geographies* within Anglophone geography – even before saying anything much about the book itself – has to do with the emphasis that I intend to place on the importance of the volume (even for those who completely rejected Soja's 'postmodern' vision) in the changing cartographies of the discipline in that tumultuous and incredibly prolific end of the decade, when postmodernism seemed to embody at once a blessing and a curse for the future of geography. It is also important to note that many of the reviews of *Postmodern Geographies* considered it together with Harvey's *The Condition of Postmodernity* (see Woodward and Jones, Chapter 15 this volume; see for example Friedland, 1992; Marden, 1992; Massey, 1991; McDowell, 1992; Warf, 1991) – although quite different in spirit, the two works were somehow perceived as marking a new opening in the reflection on postmodernism and postmodern geography.

Key arguments

The intellectual ambition and the theoretical breath of *Postmodern Geographies* was simply astonishing. After having contributed for many years to some of the most important debates on Marxism and social theory in geography (see for example Soja, 1980), *Postmodern Geographies* launched Soja on a new and provocative research agenda. In his attempt to re-establish a critical spatial perspective in/on contemporary social theory, Soja highlighted what he perceived as one of the endemic problems marking the great bulk of social inquiry from the end of the nineteenth century on: that is, *the privileging of time over space in the understanding of the socio-spatial fabric*. Soja's target then, from the opening pages of the book, was the pervasive historicism of modern social science that had, for far too long, neglected the importance of the 'spatial', allowing 'time' to dominate all interpretations of the workings of the 'social'. He suggested starting anew with a fundamental rethinking of the ontological foundations of modern European social science, in order to inaugurate a critical human geography based on new spatial ontologies and focused on what he described as 'the socio-spatial dialectic'. This grand enterprise appealed to theorists such as Lefebvre and Foucault, but also to Berger, Jameson, Giddens and many others, in order to argue for a geographical materialism centred on space, time, and being, conceived as a potential way of overcoming the 'stranglehold' of historicism that privileged *time over space*; a way to force a recognition of the deep 'spatiality of social life' and its profound political implications (Warf, 1991: 101).

The relevance for geography of the message that Soja's book sent to the rest of the social sciences can hardly be overrated. According to Soja (1989: 1), the answer that geography could offer to this problematic privileging of time over space was, in fact, the 'insistent premise and promise of postmodern geographies'. By linking geography and the spatial with the postmodern, Soja attempted not only to dismantle the implicit privileging of geometric narratives of space in the past century, but also to argue that a new geographical way of understanding society and its relations was needed – and that the return of philosophy and critical theory into the realm of human geography was the key to this revolution.

Postmodern Geographies was considered, even by its most forceful critics, a very important book for geography – and for the social sciences more broadly. Derek Gregory (1994: 258), for one, argued that '*Postmodern Geographies* [is] a brilliant book. The work of a master-craftsman, its intellectual sparkle is the product of a rare and generous critical intelligence. [...] the result is a carefully polished text, with each word weighed and set in place to bring out the deeper tonalities of the others'. Taking a brief look at the volumes that in the last ten/fifteen years have tried to reconstruct the recent history of the discipline in the English speaking world, *Postmodern Geographies* (and Soja's work more broadly) figures prominently in discussions of the postmodern turn. For instance, Cloke, Philo and Sadler's (1991) *Approaching Human Geography* highlights *Postmodern Geographies* as 'the book' that, together with Harvey's work, opened the ground for a reflection on postmodern geography: their textbook also includes, indeed, a 'window' dedicated to *Postmodern Geographies*, presented as a key text in the recent history of the discipline. Peet's seminal *Modern Geographical Thought* similarly points to the publication of *Postmodern Geographies* as a fundamental step in the affirmation of a postmodern geography and, despite some scepticism, notes that it was the first book that engaged with a 'materialist interpretation of spatiality' conceived in terms of a new 'spatialized ontology' (Peet, 1998: 222–224). *Postmodern Geographies* is also granted attention in Johnston's by now-classic disciplinary reconstruction *Geography and Geographers*, where a number of Soja's quotations open the subchapter devoted to Postmodernism (1997: 271, 275). The list is potentially endless but the important point to be made is that some of the most respected voices in the discipline have, over the years, marked out the publication of the book as a milestone in geographical reflection and a key event in the putative postmodern turn of the discipline.

Reasserting space

Postmodern Geographies consisted of a fierce challenge to historicism, to the subordination of space in critical social theory, together with an accusation of what Soja saw as the paralysis of modern geography – a paralysis due not only to this subordination, but also to the marginalization of the discipline in mainstream critical social theory. Soja's book represented, at the same time, an attempt of reaffirming the importance of the return of philosophy in geography *and* of geography in the social sciences, in those years facing the promising consequences of the spatial and cultural turns. It is probably here that we can find the book's most intriguing links/challenges to today's disciplinary reflection.

In the volume, Soja reconstructed the history of the discipline, outlining the affirmation of a historicist vision of critical thought (and the eventual banning of theory in geography, to put it in crude terms) as a consequence of the fact that the last two decades of the nineteenth century and the first two of the following were essentially dominated in Europe by 'modern movements' which defined 'separate and competitive realms of critical social theorisation, one centred in the Marxist tradition, the other on more naturalist and positivist social science' (Soja, 1989: 29). Soja urged readers to look to the *fin de siecle* – when critical theory turned resolutely historicist – to argue that, after the Paris Commune, historicity and spatiality would never again be considered *together* as explanatory forces. After the defeat of the Parisian experiment, according to Soja, the spatial dimension of critical social analysis began to fade away to give ... 'space' to the grand revolutionary Subject of History. Critical geographical readings had therefore been marginalized and silenced in social theory, with the hegemony of historical materialism (on the critical side) and scientific positivism (on the other) becoming virtually overwhelming. It is in this dialectical relationship that the reasons for the 'virtual annihilation of space by time in critical social thought' (Soja, 1989: 31) can be found; and it is also here that the exile of geography from theory formation in Western thought can be explained. The triumph of geometric spatialities (through the a-critical adoption of cartographic parameters), on the one hand, and of mere descriptive techniques, on the other, left modern geography an orphan of theory and deprived it of critical perspectives.

Seen from a 'European' perspective, Soja's discussion of twentieth-century geography, although insightful and useful, did not sufficiently take into account the evolution of continental geography and, in particular, the tensions between the *Erdkunde* project – as a critical vision of the Earth – and the emergent *state geography* that influenced most of the developments in the discipline for almost a century up until the humanistic and cultural turns. The 'spatial perspective' did not abandon geography with the turn of the century. What did occur was that a very specific cartographic vision of space emerged and colonized the discipline and social theory more broadly – for a distinct set of political reasons linked to the affirmation and consolidation of the bourgeois nation state. Such a vision would remain with us for most of the twentieth century, presenting itself as the only possible conception of space. This conception of space was metabolized in the social sciences as an essential dimension of 'reality', as a constitutive element of the world; and it is this very fact that occluded for many decades, especially in geography, any reflection on the deep nature of the cartographic reasoning on space. *Postmodern Geographies* was genuinely concerned about this lack of reflection although (in my view at

least) Soja's reconstruction somewhat overlooked the political roots of this problem: in deciding not to engage directly with some of geography's canonical texts – such as the *Erdkunde* project, Ratzel's opus, or with Vidal de la Blache's definitive 'statalisation' of the discipline – Soja did not fully confront the historical and political reasons for the dismissal of critical spatial theory in continental geography. It is curious, indeed, that geographers have made so little of this question in the discussions that followed the publication of the book, with attention mostly focused on a whole other set of issues.

Let us go back to the text, however. With an extraordinary historical leap – unfortunately typical of many Anglophone reconstructions of the discipline – *Postmodern Geographies* moved from the crisis of critical spatial theory at the turn of the century to the new geographies of the 1960s and 70s. And this is where the most fascinating and innovative part of the reconstruction begins. After having set aside the quantitative wave as a 'mathemathized version of geographical description' (Soja, 1989: 51), the discussion focused on the rise of Marxist geography in the 1970s and the 'spatializing' language adopted by an important part of Western Marxists. Soja engaged with some of the leading Marxist thinkers of the day, from Henri Lefebvre, John Berger and Ernst Mandel to Anthony Giddens, David Harvey and Frederic Jameson, while also interrogating Heidegger's spatialities and the phenomenological geographies inspired by Husserl and Sartre.

Beyond Soja's long-standing dialogue with Lefebvre's work, also of particular importance was PG's engagement with Foucauldian theories of power – a set of ideas and concepts by now taken for granted in critical human geography but revolutionary in those years. Writing in 1992, Chris Philo (together with Felix Driver, 1985, one of the pioneers in exploring the implications of Foucault's work for geography) would write: 'What is surprising is the *absence* to date of any sustained theoretical engagement with Foucault on the part of theoretically minded geographers. There is certainly one exception in this respect, though, and this is to be found in the opening chapters of Soja's important text *Postmodern Geographies*, where he examines Foucault's "ambivalent spatiality" as a straw in the wind of "*re*asserting space in critical social theory". The result is a thoughtful and intriguing introduction to Foucault's geography' (Philo, 1992: 138).

It is indeed on French social theory that Soja draws most heavily for the new mappings of Western Marxism and its passage to postmodernism, a passage that Soja presents also as a necessary reconstruction of a critical human geography. Here lies, to my mind, probably the most important contribution of this book to the discipline of geography. *Postmodern Geographies* was, in fact, about the foundation of a new ontology – of a new *spatial* ontology. A new ontology that, according to Soja, would lie at the centre of a new, 'postmodern' geography; of a new way of understanding space and society.

Postmodern geographies

Soja associated postmodernism with the process of ideology formation and cultural reproduction. Postmodernism was, primarily, a 'periodisation' that made 'theoretical and practical sense of [the] contemporary restructuring of capitalist spatiality' (Soja, 1989: 159); it was a historical transition as well as a shift in critical theory. Accordingly, contemporary geography needed to be deconstructed and subsequently reformulated in order to be able to cope with what Soja defined as the three different paths of spatialization that had marked society – and critical reflection on the social – in the preceding decades. *Postmodern Geographies* spoke of a transition in the configuration of everyday materialities, and a correspondent new set of critical

understandings; it noted the consequences of a crisis of representation that necessarily called for entirely new ways of describing our 'postmodern' reality. 'Postmodernism' was, in other words, the cultural and ideological reaction to an epochal change that Soja defined also in terms of post-historicism – the reassertion of space in Western thought – and post-Fordism – the flexible, disorganized regime of capitalist accumulation transforming the world and its political – and cultural – economies. Understanding and describing these new spatializations was key to the emergence of a new postmodern geography.

A fundamental concept in the constitution of this new geographical language was that of *socio-spatial dialectics*, based within a new understanding of 'spatiality' itself: 'the generative source for a materialist interpretation of spatiality is the recognition that spatiality is socially produced and, like society itself, exists in both substantial forms (concrete spatialities) and as a set of relations between individuals and groups, an "embodiment" and medium of social life itself' (Soja, 1989: 120). This understanding drew broadly on Lefebvrian theorizations and, inspired by the French philosopher, Soja fundamentally rejected the reduction of spatiality to the materialities of the world – as Gregory (1994: 273–274) would note, 'apprehended as a "collection of things" ("the illusion of opaqueness")' – or its reduction 'to psychological constructs, revealed as merely projections of the mind ("the illusion of transparency")' (Gregory, 1994: 273–274). Spatiality, Soja suggested, should rather be theorized as *socially produced* space. Spatiality was part of a 'second nature' (Soja, 1989: 129), at the same time source, mediator and output of social action; a spatio-temporal structuring process, argued Soja, defines 'concrete spatiality – actual human geography – [as] a competitive arena for struggles over social reproduction' while the temporality of social life 'is rooted in spatial contingency in much the same way as the spatiality of social life is rooted in temporal/historical contingency' (Soja, 1989: 130).

A truly postmodern geography would thus be a geography able to engage with the concepts of spatiality and socio-spatial dialectic in order to make sense of a new ontology based on an equally novel understanding of the relationship between space, time and being and, at the same time, of a radical urban and regional restructuring associated with the newly produced post-Fordist capitalist spatialities. It is in this sense that Soja's call for 'critical regional studies' was to be read, an approach able to provide the theoretical and methodological language for the analysis of the spatial consequences of post-historicism, post-Fordism and postmodern culture.

Los Angeles constituted, for Soja, the ideal laboratory for the exploration of the postmodern city and for the deployment of these new critical geographical tools. 'Nothing in the book prepares us for the tour de force which is Chapter Nine! Entitled "Taking Los Angeles apart: towards a postmodern geography", this is a brilliant (dis)integration of the city and the region. [...] This essay alone is worth the price of the book!!!' was Michael Dear's (1991: 651) enthusiastic assessment. 'Taking Los Angeles apart' inaugurated, indeed, an entirely new geographical language, a new way of doing geography. Soja's flight over L.A. was truly breathtaking, fuelled by powerful insights and a rare descriptive power. L.A. was, in Soja's (1989: 221) words, 'the place where it all seems to "come together": I do not mean that the experience of Los Angeles will be duplicated elsewhere', he would argue, 'but just the reverse may indeed be true, that the particular experiences of urban development and change occurring elsewhere in the world are being duplicated in Los Angeles'.

Yet what certainly became one of the most quoted bits of the book also attracted criticism. Derek Gregory's dissection of Soja's postmodern geographies in his influential *Geographical Imaginations* (1994, see Pickles, Chapter 20 this volume) (which dedicates an entire chapter to the book) lamented that in Soja's 'spiralling tour around the city' (Gregory, 1994: 299) his postmodern geography 'as a whole seems indifferent to the importance of ethnography and in particular to the experiments in "polyphony" which have so vigorously animated postmodern ethnography' (Gregory, 1994: 295). This was in part a consequence, according to Gregory, of the visual metaphors through which Soja conceived spatiality; 'his representation of the landscapes of LA is conducted from a series of almost Archimedian points from which he contemplates a series of abstract geometries' (Gregory, 1994: 299). Gregory's critique was motivated by the fact that, despite Soja's explicit recognition that any totalizing description of the postmodern city (here, L.A.) would have been impossible (would, indeed, represent a contradiction in terms within the framework of a 'postmodern geography') and thus his empirical chapters could only consist of a fascinating succession of fragmentary glimpses, the excursion presented in Chapter Nine was, nonetheless, conducted from a series of external (albeit mobile) viewpoints. This criticism, in many ways not surprising if we consider the experimental nature of the chapter, should in no way diminish the path-breaking impact of those 'flights'; the fact that Soja's 'spiralling tours' of Los Angeles created an entirely new and provocative way of writing geography.

The most serious accusations against PG were levelled, however, at its silence with respect to feminist theorizations of space and society. 'The most startling absence from PG is any sustained engagement with feminism. This is perfectly consistent with the modernist cast of Soja's prosject as a whole, I suppose, and a number of writers have drawn attention to the masculinity of the modern gaze and its systematic devaluation of the experience of women', Gregory (1994: 309) underlined, noting also that 'the contemporary conjunction between postmodernism and feminism makes such an omission from a postmodern geography truly astonishing'. Criticism was hence forthcoming from many feminist geographers (see, for example, Deutsche, 1991; McDowell, 1992) but it was Doreen Massey's influential review article 'Flexible Sexism' that set the tone of the debate. Launching a forceful attack on Soja's and David Harvey's 'postmodern' works, it accused both authors of essentially ignoring the feminist contribution to spatial and social theory: 'it seems to me that the absence from, indeed the denial by, both these books of feminism and the contributions it has recently made, raises issues which are important for all of us, and which range from our style as academics to the way in which some of the central concepts of the debate are formulated' (Deutsche, 1991: 31–32) (see also Woodward and Jones, Chapter 15 this volume). It is important to note that Soja would, by and large, accept this criticism and, in the opening pages of his subsequent book, *Thirdspace*, would admit to what could have been perceived as a gender-bias in his previous book, albeit rejecting the idea that PG 'can be read as a masculinist posing tout court' (Soja, 1996: 13). Indeed, *Thirdspace* would extensively engage with feminist and postcolonial theory, with particular attention, in geography, to the work of Gillian Rose (Soja, 1996: 121).

If the controversy over feminism and postmodernism marked an important passage in the reception of the book, another aspect highlighted by many commentators was the 'thoroughly *modernist*' spirit of Soja's project (Gregory, 1994: 273). Warf (1991: 101), suggested, for example, that 'Soja's explanation amounts to an impressive historical-materialism interpretation of the late twentieth century capitalism that hinges upon a decidedly modernist, not a postmodernist view', while for Dear (1991: 653), Soja's Marxism effectively left the 'ontological project' intact, overvaluing the contribution of Marxist mainstream, and representing most non-Marxists as fundamentally anti-postmodern.

What do we make of this criticism today? My view, after almost two decades, is that *Postmodern Geographies* was in fact *intentionally* conceived as a Marxist and modernist critical rethinking of geography and of the hegemony of historical materialism in the social sciences. Only with the last chapter on L.A. did Soja explicitly move beyond that tradition, experimenting with what a postmodern geography of a postmodern city could be. However we may interpret the success of that experiment, I do believe that this was the 'substance' of the project behind PG. We should also recognize that the postmodern overall had entered Anglophone geography largely by way of an intricate but decidedly Marxist stream, and this can explain why the tension between a declared postmodern relativism and structurally oriented 'Marxist' descriptions of society and the economy marked (postmodern) geography for so long. The emphasis placed by Soja on the transformation of global modes of capitalist production is also understandable from this perspective.

A similar point can also be made regarding a more general reflection on the potential political implications of the sort of postmodern praxis described by Soja. The 1990s witnessed a rich debate on radical 'postmodern' politics and, looking back, it is important to note the place of Soja's subsequent two books *vis-à-vis* these discussions. Despite the fact that, again from my point of view, *Postmodern Geographies* was a decidedly 'political' text – it would be quite difficult to argue otherwise considering its driving ideological push – it is important to note that in *Thirdspace* first and *Postmetropolis* later, Soja would explicitly engage with many of the early critiques, such as that of having partially overlooked the politics of racism and sexism, or of not having fully explored, in his postmodern politics of resistance, the potential of many grassroots social movements emerging in those times as new political subjects.

Conclusion

Today, Soja's influence on contemporary understandings of the urban is still strong and reaches a genuinely international audience well beyond the discipline of geography. Indeed, one of the hallmarks of *Postmodern Geographies* was its truly global reach. To my knowledge, the book was translated into Portuguese, Japanese, Chinese and Serbo-Croat, and parts of it into Italian and German; a special edition has also been published in India. Copies of *Postmodern Geographies* can be found not only in academic bookshops around the world, but also in the book collections of modern art and architectural museums and galleries from Washington to Berlin to Shanghai.

What can we say of the legacy of *Postmodern Geographies* at a time in which the echoes of the postmodern wave in geography (and beyond) seem to have faded away, as evidenced by David Ley (2003) in a recent appraisal? The first thing that I would like to say is that although it might be true, as Dear (1991: 654) would have it, that *Postmodern Geographies* was 'a brilliant and fruitful account of one person's encounter with postmodernism', *Postmodern Geographies* was – and is – also a book that marks a much longer trajectory, traversing the history of human geography and of the social sciences of the last two decades. For this reason, I believe that today we are in the position to be able to express a serene and partially detached (time helps) judgement on this extraordinary initiative. The road opened by *Postmodern Geographies*, with its powerful insights and brilliant theoretical engagements, but also with all its tensions and provocations, has given a vital contribution to making geography the vibrant discipline that it is today, if anything for the reactions that it provoked and the debate that it stimulated, both in the academic and intellectual realms.

We can probably also accept today, with similar peace of mind, that *Postmodern Geographies* was perhaps not quite *yet* as postmodern as the title intimated, but it is out of the question that the volume represented a path-breaking intervention at a moment in which there was a need for just such an intervention – also for those who did not like the book and dismissed it. And here I do not refer only to Soja's early engagement with Foucault, or to his reflection on Lefebvre's spatialities or to the concept of sociospatial dialectics that will influence a generation of urban geographers. Nor do I only think of his timely recognition of the utility of Giddens' reflection for geographers and of Jameson's work for all scholars concerned with the contemporary city. *Postmodern Geographies*'s contribution has not only been important for the revolution that it brought to the reflection on the relationship between Western Marxism, postmodernism and geography, but also for its fantastic flight over L.A. that launched a new way of writing geography. What I believe has been the major contribution of this book, especially in the light of what happened in the decades that followed, is that it showed how geography might (and ought) to dialogue with philosophy and social theory. What is more, as a geographer, Soja demonstrated that in order to grasp the spirit of our times, to interpret our new modes of communication, to read – and live – the cities of today, a geographical way of understanding space and spatial theory was not only useful but crucial for the future of social analysis. 'Soja has irrevocably shifted the perspective of both geographers and nongeographers on the project of Geography' – announced Dear in the by-now far away 1991; contemporary human geography has shown that this was a fitting prediction.

Secondary sources and references

Cloke, P., Philo, C. and Sadler, D. (1991) *Approaching Human Geography*. New York: Guilford.

Dear, M. (1991) 'Review of Postmodern Geographies', *Annals of the Association of American Geographers* 80: 649–654.

Deutsche, R. (1991) 'Boys Town', *Environment and Planning D: Society and Space* 9: 5–30.

Driver, F. (1985) 'Power, space, and the body: a critical assessment of Foucault's "Discipline and Punish"', *Environment and Planning D: Society and Space* 3: 425–446.

Eflin, J. (1990) 'Review of Postmodern Geographies', *Geographical Review* 80: 448–450.

Farinelli, F. (1992) *I segni del mondo*. La Nuova Italia Scientifica: Bari.

Friedland, R. (1992) Space, place, and modernity: the geographical moment, *Contemporary Sociology* 21: 11–15.

Gregory, D. (1994) *Geographical Imaginations*. Oxford: Blackwell.

Harvey, D. (1989) *The Condition of Postmodernity*. Oxford: Blackwell.

Johnston, R. (1997) *Geography and Geographers*. London: Arnold.

Ley, D. (2003) 'Forgetting Postmodernism? Recuperating a social history of local knowledge', *Progress in Human Geography* 27: 537–560.

McDowell, L. (1992) 'Multiple voices: Speaking from inside and outside "the Project"', *Antipode* 24: 56–72.

MacLaughlin, J. (1994) 'Review of postmodern geographies', *History of European Ideas* 18: 803–805.

Marden, P. (1992) 'The deconstructionist tendencies of postmodern geographies: a compelling logic', *Progress in Human Geography* 16: 41–57.

Massey, D. (1991) 'Flexible Sexism', *Environment and Planning D: Society and Space* 9: 31–57.

Peet, R. (1998) *Modern Geographical Thought*. Oxford: Blackwell.

Philo, C. (1992) 'Foucault's Geography', *Environment and Planning D: Society and Space* 10: 137–161.

Resch, R. (1992) 'Review of Postmodern Geographies', *Theory and Society* 21: 145–154.

Rose, G. (1991) 'Review of Postmodern Geographies', *Journal of Historical Geography* 17: 118–121.

Smith, D.M. (1991) 'Review of Postmodern Geographies', *Urban Geography* 12: 93–95.

Soja, E. (1989) *Postmodern Geographies*. London: Verso.

Soja, E. (1980) 'The socio-spatial dialectic', *Annals of the Association of American Geographers* 70: 255–272.

Soja, E. (1996) *Thirdspace*. Oxford: Blackwell.

Soja, E. (2000) *Postmetropolis*. Oxford: Blackwell.

Warf, B. (1991) 'Review of Postmodern Geographies and The Condition of Postmodernity', *Journal of Regional Science* 31: 100–102.

Warf, B. (1993) 'Postmodernism and the localities debate: ontological questions and epistemological implications', *Tijdschrift voor Econ. en Soc. Geografie* 84: 163–168.

17 THE CAPITALIST IMPERATIVE (1989): MICHAEL STORPER AND RICHARD WALKER

Neil Coe

Most existing treatments of urbanization, regional development, and industry location are based upon neoclassical economics and share its assumptions and shortcomings. The three fundamental building blocks of neoclassical theory are: (1) the central economic activity is exchange; (2) the goal of economic exchange is efficient resource allocation in the service of subjective preferences; (3) the natural state of the system is to come to rest at a stable equilibrium ... Our view of capitalist reality is ... quite otherwise. The economy is fundamentally a disequilibrium system, driven to grow and to change by its own internal rules of surplus generation, by investment to expand capital, by fierce competition, and by technological change to extract more surplus (value) from human labor. (Storper and Walker, 1989: 38)

Introduction

The 1989 publication of Michael Storper and Richard Walker's book *The Capitalist Imperative* was undoubtedly a key milestone in the development of a Marxist-influenced political economy approach to human geography. Since the mid-1970s, and prompted by David Harvey's *Social Justice and the City* (1973), many human geographers, and economic geographers in particular, had been striving to shrug off the legacy of neoclassical location theory and its sometimes simplistic attempts to quantitatively model the spatial

structures of the economy (e.g. Isard, 1956). Geographers increasingly began to challenge the presumed neutrality of spatial science, highlighting, for example, the complete absence of capitalist class relations in such analyses. Instead, and inspired by the insights offered by Marxian political economy, they sought to demonstrate how the processes of capitalist accumulation and their associated social relations were inherently geographical. In such analyses, the main foci of analysis were no longer discrete spatial coordinates such as firm locations, but rather the deep structures of capitalist social relations – such as the ongoing struggle between capital and labour – that were in turn seen to actively shape spatial patterns of economic development.

A wide range of influential books and articles were published by geographers in the late 1970s and 1980s as part of this drive to develop detailed historical geographies of the capitalist mode of production and in so doing forge a 'new' industrial geography to supplant the previously dominant neoclassical paradigm. Scott (2000) identified three strands of research within this emerging body of work. One theme focused on the nature of urban space under capitalism, and more specifically, the links between land rent, housing provision and urban planning (e.g. Harvey, 1985). Another was concerned with problems of deindustrialization, poverty and job loss occurring as economic restructuring bit hard in many developed countries (e.g. Bluestone and Harrison, 1982) and was closely linked to detailed investigations of regional industrial restructuring and changing geographies of labour (e.g. Massey, 1984). The third stream of work sought to theorize uneven geographic development within capitalism at a range of spatial scales (e.g. Harvey, 1982; Smith 1984; see Castree, Chapter 8 this volume; Phillips, Chapter 9 this volume). As we shall see, Storper and Walker's *The Capitalist Imperative* – which offers 'a broad depiction of the complex interconnections between technology, industrialization and territorial change in advanced capitalist societies' (Smith, 1992: 81) – can be positioned as an ambitious attempt to bridge the latter two strands of political-economic research.

The book and its argument

The argument of the book can be summarized, albeit somewhat crudely, as follows. The geographies of capitalist development are extremely dynamic, with the fortunes of particular places, regions and countries waxing and waning over time as they interact with different waves of industrialization. *Contra* the assumptions of neoclassical economics and location theory, rather than exhibiting any tendency to spatial uniformity, capitalist growth is inherently expansionary and unstable, and produces differentiation between territories specializing in particular industries. While industries exhibit an innate tendency to agglomerate in particular localities, the tapestry of interconnected clusters that results is far from static, and the production process, rather than the market, is central to understanding this dynamism.

The most important shapers of industrial change are technologically intensive industries with innovative new products and technologies which have the ability to re-locate and actively create new regional clusters during the early stages of their development. The key

mode of economic organization within this fluid system is not the market, the individual firm or the workplace; rather it is the territorial production complex, a cost-reducing confluence of firms, workers, knowledge, resources and infrastructure (see below). Such complexes facilitate the flexibility – in terms of divisions of labour, local labour markets and technology – inherent to modern production systems. Understanding the fluctuating yet interconnected fortunes of these territorial complexes, both new and old, in the core and in new growth peripheries, is critical to revealing the geographical nature of capitalist growth and its wider impacts upon society and class relations.

According to its authors, the central aim of the book was to grapple with two theoretical challenges (Storper and Walker, 1989: 1). First, it sought to develop an analytical framework for understanding the geography of economic development, and more specifically, the geography of industrialization. As the following quote suggests, the objective here was to move well beyond the limitations of the neoclassical approach: 'We propose to thoroughly rewrite location theory from the standpoint of political economy, and, in so doing, to leave location theory behind in favour of a theory of geographical and territorial industrialization' (Storper and Walker, 1989: 3). The book sought to do so through extensively critiquing existing theories as well as developing the authors' own theorization. In addition to neoclassical models of industrial location, several other perspectives were put to the sword, including regional science, industry/sector studies, regulation theory, behaviouralism and systems theory. While in Gertler's (1991: 363) words 'these theories are lined up and shot down like so many empty bottles in a shooting gallery', the authors were also 'smart enough to borrow ideas when they make a useful contribution to their own framework'. In forging a new analytical framework out of this extensive critical review and synthesis, a specific aim was to bring together hitherto largely separate theories of industrial location and regional development.

It is worth briefly reviewing the most important analytical concepts proposed and developed by Storper and Walker in their account:

- *Geographical industrialization* describes the inherently geographical nature of the growth and decline of capitalist industries and the constantly shifting map of economic activity that results. There are four key stages: (1) new industries develop in particular places; (2) similar and related activities cluster around this new 'hearth'; (3) industries disperse some of their production capacity away from these clusters, and (4) new industries, based on innovative products and technologies, shift away and take up new locations.
- *Territorial development* captures these processes from the perspective of places. It speaks to the ways in which territories – be they cities, regions or countries – develop through successive waves of industrialization. Over time, new places are connected into the capitalist system, while others based on declining industries will become less significant. Territorial development is heavily shaped by the 'dominant ensemble' of industries at a given point in time, such as automobile and consumer durables production in the middle decades of the twentieth century.
- *Windows of locational opportunity* are moments of locational freedom during the early stages of development of technologically innovative industries (i.e. before they have invested heavily in factories, etc.) in which leading firms can choose between a range of viable locations. This idea is underpinned by a recognition that industries do not simply respond to the location of inputs such as labour, resource inputs and suppliers, but that they also *create* and *attract* such factor inputs.
- *Territorial production complexes* such as cities, or city-regions, are seen as the main mode of organizing capitalist production rather than firms, markets or workplaces. They represent localized arenas in which costs for participating firms and industries are lowered through the development of an effective division of labour between and within firms, the sharing of infrastructure, the pooling of resources, the development of labour skills, and the creation of an environment conducive for innovation. Over time, however, locating in such a complex may start to constrain an industry and leave it vulnerable to approaches developed elsewhere.

Second, it strove to contribute to social theory more broadly by demonstrating how political economic processes are profoundly shaped by geography. In that sense it can be read as part of a wider project to reveal the mutually constitutive interactions between the social and spatial structures of society (e.g. Gregory and Urry, 1985). As Storper and Walker argue, 'Space must again become an active variable in the social system, because human action takes place in specific locales and social relations from both within and across territorial boundaries … That is human life unfolds in a thoroughly geographical way' (Storper and Walker, 1989: 3). The idea, then, was to fashion a form of spatial-temporal analysis that would make an impact across the social sciences, and not just be read by those interested in cities, industries and economic change.

Placing the book and its authors in context

In attempting these twins tasks, *The Capitalist Imperative* was clearly influenced by two inter-related theoretical developments in the social sciences more generally during the 1980s, namely those concerning critical realism and the structure-agency debate. Philosophically, the book is underpinned by critical realism, a theory developed in the geographical context by Sayer (1984) and which gained increasing currency during the 1980s as scholars sought to break down the certainties and economic logics of structural Marxist frameworks. By contrast, realism sought to emphasize the contingent factors that determine how the causal relationships inherent to capitalism actually play out (differentially) on the ground. Such an approach is implicit in Storper and Walker's book: 'the narrative is replete with "mights", "maybes", and "possiblys", though one is never allowed to lose sight of the necessary conditions and the underlying structures through which basic causal properties are defined. The overall effect is of a sensitivity to the determinants of the concrete rich enough to satisfy the most hardened of critics of so-called Marxist reductionism' (Cox, 1991: 442). Similarly, it is also possible to discern echoes of Gidden's (1984) 'structuration' theories, which sought to map out a middle ground between those seeking to emphasize either the importance of structural forces or human agency (Gertler, 1991). These underlying philosophical positions are made clear when the authors state how they 'try to walk the fine lines between various intellectual pitfalls. We recognize both the force of social structure and the initiatives of human agency; both necessity and contingency in the causation of particular events; both the widespread impact of deep structure and the relative autonomy of various parts of society…' (Storper and Walker, 1989: 5).

Returning to the different lines of political economic enquiry identified earlier, *The Capitalist Imperative* operates at a level of abstraction that falls between that of Harvey's (1982) *The Limits to Capital* or Smith's (1984) *Uneven Development* as compared to Massey's (1984) *Spatial Divisions of Labour* (see Phelps, Chapter 10 this volume), all three of which are key texts in geography's political economy canon. The former two volumes present overarching theoretical models of the spatial constitution of capitalism, and are rich in analytical insight, but thin on empirics. Massey's text, in comparison, presents an empirically strong account of regional restructuring in the UK in the 1960s and 1970s in developing her notion of spatial divisions of labour (interestingly, writing in 2001, one of the authors reflected on the importance of Massey's book along with Bluestone and Harrison's, 1982, treatise on American deindustrialization as key intellectual 'jumping-off points' for the work which led to *The Capitalist Imperative* – see Walker, 2001). Storper and Walker offer an account that is largely theoretical in nature, but one which describes the shifting position of different kinds of places within the capitalist system, and provides selective, rather than comprehensive, empirical evidence to support its arguments. As they argue in the last few lines of their book, their approach to the 'inconstant geography' of capitalism is one that 'can get hold on historical change at a middle level of abstraction and concreteness, between the full sweep of centuries and the endless swirl of everyday events' (Storper and Walker, 1989: 227).

A final way to position and think about this book is as a product of a particular time and place. Storper and Walker's book was part of a wider corpus of path-breaking work by a 'Californian School' of economic geographers – which also included Amy Glasmeier, Ann Markusen, AnnaLee Saxenian, Erica Schoenberger and Allen Scott, among others – who were using the experiences of California and its growth industries to inform wider debates about the resurgence of regional economies in a post-Fordist era. This context gives the book a certain 'flavour' and perspective, as recognized by the authors themselves in the Preface when they state: 'Our perspective has been skewed in important ways by the view from California, on the edge of the booming northern Pacific Rim; hence the overriding emphasis given here to economic growth and geographic expansion in modern industrialization ... others are better positioned to depict the devastation of unemployment in the First World or underdevelopment and imperialism in the Third' (Storper and Walker, 1989: ix).

The authors of the book, both Associate Professors at the time of writing (Storper at the University of California, Los Angeles – UCLA – and Walker at the University of California, Berkeley), have gone on to full professorships and have firmly established themselves as leading figures in economic geography (although their influence stretches well beyond the subdiscipline). Walker has remained at University of California, Berkeley, and since the early 1990s has turned his attention to analysing the historical economic development of California. In his work and writing, he has retained a strong belief in the power of the political economy perspective (e.g. see Walker, 1998). Key subsequent publications include his co-authored book with Andrew Sayer developing the notion of the social division of labour (Sayer and Walker, 1992), and latterly, a book on Californian agribusiness (Walker, 2004). During the 1990s, Storper moved away from purely economic explanations of regional dynamism and pioneered the development of more socio-cultural explanations based on knowledge, learning and the emergence of place-specific economic conventions and practices (e.g. Storper, 1997). Interestingly, Storper himself has been critical of what he sees as some of the excesses of the 'cultural turn' in geography and the social sciences and has more recently been working at the interface of geography and economics (e.g. Storper, 2001; Storper and Venables, 2004: see also Reimer, 2004).

Initial reception and critiques

This is a book that all human geographers should read ... Michael Storper and Richard Walker achieve better than any before them the goal of demonstrating how capitalism is (and can only be) spatially constituted. In doing so, they advance the intellectual project and cause of our discipline like few others books in geography before this one. (Gertler, 1991: 361)

This is a work that *must* be read by those with interests in industrial geography and regional development, but one that also has important implications for a much wider field of scholarship. (Wood, 1993: 81, emphasis in original)

Surveying the book reviews published in the few years after publication, and as the above two quotes suggest, *The Capitalist Imperative* was widely (and rightly) lauded as an impressive work not only by economic geographers, but also by historical and political geographers, and sociologists and economic historians. Commentators admired the clarity and readability of the book, and the quality and extent of the critique of existing perspectives. In terms of the own conceptual framework, there was praise, for example, for its coherence and elegance, its ability to operate at a mid-level of abstraction, the way in which it successfully drew together the industrial location and regional development approaches, its revealing of the social dimensions of labour, technology, investment and the like, and the manner in which it successfully incorporated space and uneven development into the conceptual apparatus.

That being said, a book of this theoretical breadth and ambition was not received without some misgivings. We can group these concerns into six sets. Firstly, the critique of existing approaches was seen by some to be overdone, attacking what one reviewer described as a 'cartoon version of neoclassical theory' (Solot, 1990: 359), and one which many economists of the time would themselves have seen as simplistic and outmoded. Secondly, for some readers, the authors were more effective in synthesizing and surveying existing approaches than in advancing their own approach. Smith (1992), for example, suggested that the regional mosaic they depicted did not reveal the structural 'geometries' of development and underdevelopment at different scales that characterize the world economy. This relative silence on the 'scaling up' of their analysis is problematic in the contemporary era of neoliberal globalization. A third and particular area of concern was that Storper and Walker offered no systematic attempt to sustain their theoretical arguments through empirical evidence. Several reviewers noted that the empirical evidence was highly selective both sectorally and geographically, drawing in particular on developments in certain US manufacturing sectors. The empirical evidence was also rather piecemeal: the book contains a lot of small examples rather than large sets of quantitative data or long sustained histories of particular places or sectors. As such, it was not able to offer explicit information as to why particular regions went through certain changes at particular times. Fourthly, particular elements of Storper and Walker's conceptual framework also came under the microscope. The notion that industries create regions was seen by some as too unidirectional and lacking appreciation of the two-way relationships (i.e. regions also create industries) seen in Massey's work on spatial divisions of labour. Others saw the analysis as too technologically deterministic – although the authors described their approach as 'strong technological determination' rather than determinism (Storper and Walker, 1989: 124) – and as placing too much emphasis on a version of the product life cycle, with a tendency to focus only on certain kinds of post-Fordist, flexible production. Fifthly, it was contended there were also some important 'missing links' discernible in the book. In particular, there is little room for the nation state (in its different scalar formations) and civil society in the analysis.

Relatedly, the US-centrism inherent to the evidence base (see above) is reflected in the lack of discussion of different national variations in capitalist processes shaped by institutional, regulatory and political forces. Finally, the book does not explicitly draw out the policy implications that stem from its analysis of contemporary capitalist growth: one reviewer lamented the apparent 'steadfast resistance' to engaging in policy debates (Gertler, 1991: 364).

Looking back: a lasting legacy?

Assessing the lasting impact of any intellectual contribution is necessarily a highly qualitative and subjective exercise. However, quantitative indices such as citation indexes can provide a starting point for such as assessment. According to the *ISI Web of Science*, as of September 2006 the book had been cited 520 times in articles published in listed journals. From 1992 until 2002, the book received at least 30 cites per year, and still currently receives around 20 per year, and that is just citations in leading journals, ignoring citations in books and myriad other forms of publications. By this simple measure alone, then, the ongoing significance of this book starts to become clear.

Notwithstanding some of the critiques raised above, the book has arguably had a particularly strong impact of three inter-linked strands of economic geography research:

• *Post-Fordist production systems*: Storper and Walker's volume was a key contribution to the aforementioned 'Californian School' of research which theorized the agglomerative tendencies inherent in the organizational and technological dynamics of a range of post-Fordist craft, high technology and business service industries, and in turn precipitated a lively debate within geography about the validity and wider applicability of these ideas.

- *New regionalism/geographies of innovation*: relatedly, through its emphasis on labour, technology and investment as social processes, and innovations as 'placed' in particular localities, *The Capitalist Imperative* also helped sow the seeds for the policy-oriented 'new regionalism' agendas that emerged from the mid–1990s. These debates have sought to recast regional development in an era of globalization in terms of differing combinations of localized production cultures and conventions, institutional and labour market formations, and dynamic learning and innovation processes.
- *Labour geographies*: the book depicts how labour forces are 'made' in territorial production complexes through a variety of social processes, and shows that labour is much more than simply a passive factor of production. Instead, local labour markets are constructed, segmented and altered through the location and relocation of production. These insights were an important precursor to the now small but flourishing subdisciplinary area of labour geographies (e.g. Herod, 2001).

Overall, perhaps the most important contribution of the book is the forging of a coherent and integrated theory of capitalist uneven development that is 'grounded' enough to deal with the volatility, variability and contingencies of real world economies.

Conclusion

The economic geography of today is very different from that of 1989. Influenced by the cultural turn, the latest 'new' economic geography is a theoretically and methodologically plural and diverse field, ranging from quantitative approaches through to post-structural accounts grounded in textual and discourse analysis. For those familiar with recent developments in the field, *The Capitalist Imperative* may seem to be lacking in several important respects. It may seem rather structuralist, with the agency of places and actors rather circumscribed by broader capitalist dynamics. It may seem overly economistic, with more cultural processes and formations underweighted. It may seem overly productionist, with little or no mention of retail capital and processes of consumption, for example. It may appear rather technologically deterministic. And the institutional landscape on which events are played out might appear rather 'thin', with the absence of different states-formations particularly notable. There are elements of truth to all these critiques, and indeed the authors have themselves clarified and in some cases modified their positions on these issues in subsequent writings. On the other hand, these critiques are in some ways unfair, as any book is a product of its particular time (and place). *The Capitalist Imperative* undoubtedly drove several debates in economic geography forward productively, and it is in those terms that it should be judged.

As a student reader, a useful test of your own might be to focus on some of the key economic-geographical developments of today – the rise of India and China as economic superpowers, for example – and think as you read about the extent to which the book offers a framework for understanding those processes. You will find that the book can reveal a lot about some aspects – the development of new manufacturing zones and the technologies and labour market conditions that underpin them, for example – but less about others, such as the importance of Western retailers and consumption patterns in driving demand for goods, and the role of the Indian and Chinese states in shaping the pace and nature and development within their countries. What reading *The Capitalist Imperative* will make clear, however, is that a coherent geographical-political-economic perspective must be a central part of our attempts to understand such phenomena.

Secondary sources and references

Bluestone, B. and Harrison, B. (1982) *The Deindustrialization of America*. New York: Basic Books.

Cox, K. (1991) 'The Capitalist Imperative' (book review), *Political Geography Quarterly* 10 (4): 441–444.

Gertler, M.S. (1991) 'The Capitalist Imperative (book review)', *Economic Geography* 67 (4): 361–364.

Giddens, A. (1984) *The Constitution of Society*. Cambridge: Polity Press.

Gregory, D. and Urry, J. (1985) (eds) *Social Relations and Spatial Strucures*. London: Macmillan.

Harvey, D. (1973) *Social Justice and the City*. London: Arnold.

Harvey, D. (1982) *The Limits to Capital*. Oxford: Blackwell.

Harvey, D. (1985) *The Urbanization of Capital*. Oxford: Blackwell.

Herod, A. (2001) *Labor Geographies*. New York: Guilford.

Isard, W. (1956) *Location and Space Economy*. New York: Wiley.

Massey, D. (1984) *Spatial Divisions of Labour: Social Structures and the Geography of Production*. London: Macmillan.

Reimer, S. (2004) 'Michael Storper', in P. Hubbard, R. Kitchin and G. Valentine (eds) *Key Thinkers on Space and Place*. London: Sage, pp. 282–287.

Sayer, A. (1984*) Method in Social Science*. London: Hutchinson.

Sayer, A. and Walker, R. (1992) *The New Social Economy: Reworking the Division of Labour*. Oxford: Blackwell.

Scott, A.J. (2000) 'Economic geography: The great half-century', *Cambridge Journal of Economics* 24: 483–504.

Smith, N. (1984) *Uneven Development: Nature, Capital and the Production of Space*. Oxford: Blackwell.

Smith, N. (1992) 'The Capitalist Imperative (book review)', *Antipode* 24 (1): 81–83.

Solot, M. (1990) 'The Capitalist Imperative (book review)', *Journal of Historical Geography* 16 (3): 358–359.

Storper, M. and Walker, R. (1989) *The Capitalist Imperative: Territority, Technology and Industrial Growth*. Oxford: Blackwell.

Storper, M. (1997) *The Regional World: Territorial Development in a Global Economy*. New York: Guilford Press.

Storper, M. (2001) 'The poverty of radical theory today: From the false promises of Marxism to the mirage of the cultural turn', *International Journal of Urban and Regional Research* 25: 155–179.

Storper, M. and Venables, A.J. (2004) 'Buzz: Face-to-face contact and the urban economy', *Journal of Economic Geography* 4: 351–370.

Walker, R. (1998) 'Foreword', in A. Herod, (ed.), *Organizing the Landscape: Geographical Perspectives on Labor Unionism*. University of Minnesota Press, Minneapolis, pp. xi–xvii.

Walker, R. (2001) 'Bennett Harrison: A life worth living', *Antipode* 33 (1) 34–38.

Walker, R. (2004) *The Conquest of Bread: 150 years of California Agribusiness*. New York: The New Press.

Wood, A. (1993) 'The Capitalist Imperative (review essay)', *Antipode* 25 (1): 69–82.

18 THE GEOGRAPHICAL TRADITION (1992): DAVID LIVINGSTONE

Nick Spedding

> Stories are always told by people, about people, for people. Geography's story is no exception. (Livingstone, 1992: 4)

Introduction

First published in 1992, David Livingstone's *The Geographical Tradition* is widely regarded by English-speaking geographers as the most important history of the discipline produced in recent years. The ubiquity of *The Geographical Tradition* on reading lists for courses on the history and philosophy of geography, and its rapid appearance in the journal *Progress in Human Geography*'s series on 'classics revisited', testify to this (Mayhew *et al.*, 2004). If the subject matter of the book was mostly familiar – *The Geographical Tradition* examines many of the same people and ideas as previous histories of geography – the way in which Livingstone told his story was not. Because it is a remarkably open discipline – a core concern with the spaces and places of the earth's surface potentially includes far more than it excludes – it is not surprising that geography's past contains a huge variety of phenomena studied with a huge variety of tools for thinking and doing. This substantive, philosophical and methodological diversity was recognized by historians of geography long before *The Geographical Tradition* was published. The book made a big impact because it did far more than just list this past variety; Livingstone set out the detail of *how* and *why* such variety arose. In particular, *The Geographical Tradition* destroyed the notion – implicit, if not explicit, in many previous accounts – of smooth progress towards an ideal, objective truth. For Livingstone, academic geography must be a subjective, plural, *contested* enterprise because it is always influenced by the particular times and places in which geographers work:

> Geography has meant different things to different people in different settings ... the heart of my argument is that geography changes as society changes, and that the best way to understand the tradition to which geographers belong is to get a handle on the different social and intellectual environments within which geography has been practised. (Livingstone, 1992: 347)

The inspiration for Livingstone's *contextual* approach to geography's history came from studies in the history and philosophy of science, especially those that discussed the sociology of scientific knowledge. From these came *The Geographical Tradition*'s insistence that geographical knowledge was, and is, always *partial*, in the sense that it is neither value-free nor complete. In turn, this overt scepticism towards the possibility of absolute truth aligned Livingstone's history with aspects of postmodernism. Postmodern thinking was a powerful influence on much geographical research in the 1990s, and continues to be so (see Minca, Chapter 16 this volume; Woodward and Jones, Chapter 15 this volume). This, combined with the quality of Livingstone's scholarship, does much to explain why *The Geographical Tradition*'s reputation has diminished little over time.

The author

Based in the School of Geography at Queen's University in Belfast, Northern Ireland, David Livingstone has worked his way up through the academic ranks, and is now Professor of Geography and Intellectual History. In the Royal Geographical Society's Biographies of Chartered Geographers, Livingstone summarizes his research interests as 'history and theory of geography, cartography and scientific cultures'. He is recognized in academic circles as both a geographer and a historian of science. It is important to appreciate this twin identity to understand the scope and substance of Livingstone's work, and the high regard in which he is held by his peers. His consistent focus on the links between geography and the wider sciences, arts and humanities means that he is far less concerned to work within orthodox (but often arbitrary) disciplinary boundaries than other historians of geography have been. Before *The Geographical Tradition* was written, Livingstone's publications in the 1980s addressed such issues as the shortcomings of traditional scientific method, the use of myth, metaphors and models in science, the absence of an essential identity for academic geography and the interaction of magic, theology and science. He is an expert on Nathaniel Southgate Shaler, the nineteenth-century Harvard University polymath, and the influence of the neo-Lamarckian version of evolutionary thought on the environmental and social sciences. The common factor that unites these themes, all of which recur in *The Geographical Tradition*, is suspicion of the core

qualities – progress, certainty, objectivity – traditionally claimed for modern science.

The text

Livingstone's (1992) history of geography starts with a study of geography's historiography. *The Geographical Tradition*'s first chapter is a sustained critique of the ways in which geographers typically have tried to write the history of their discipline. The title of the chapter – *Should the history of geography be X-rated?* – was borrowed from a 1974 paper in the journal *Science* by the US historian of science S.G. Brush, entitled 'Should the history of science be rated X?' The orthodox view of scientific method, associated with positivist empiricism, emphasizes particular characteristics – measurement, rigour, logic, objectivity, etc. – that guarantee reliable, truthful knowledge. Students are usually expected to absorb and adhere to this ideal as part of their professional training. Brush's paper highlighted new work by historians of science that undermined the orthodoxy by documenting just how subjective (i.e. dependent on the thoughts and actions of the scientists themselves) real-world science was; hence Brush's mischievous suggestion that 'old school' scientists should introduce censorship to prevent impressionable students from getting hold of such seditious literature. Livingstone's point here was that there was no need to label existing histories of geography with an X-certificate because geography's historians had yet to pick up on the new ways of writing about science. The 'sanitized' histories of the discipline available, bereft of 'social context, metaphysical assumptions, professional aspirations or ideological alliances' (Livingstone, 1992: 2), were far more likely to send students to sleep than to the barricades. By writing *The Geographical Tradition*, Livingstone intended to introduce geographers to revisionist historiographies of science and to illustrate the possibilities of these with case studies selected from geography's past. Henceforth, uncritical faith in philosophical and methodological rigour guiding the impartial hero in his quest to discover indisputable truth was out; emphasis on the various human, or, to put it bluntly, 'non-scientific' factors was in. Historians of geography now had to make room for people's luck, misfortunes, whims, mistakes, friendships, rivalries, prejudices and vested interests.

Livingstone (1992) proceeds with a detailed critique of the 'standard' textbook histories. His most basic objection is that these are boring – tedious lists of who said what, when. But he argues for a greater fault – texts that present a distorted history because of their *presentism* and *internalism*. Presentist history is history written to match present-day standards, without taking account of how things were different in the past. Internalist history is history written to match a narrow, often arbitrary, view of how things should be, ignoring alternative perspectives or factors that sit outside the preferred view. Livingstone's (1992: 4) concern is that these prejudices combine to produce *Whiggish* histories, 'written backwards'. These accounts start with a particular view of *the* correct way to do geography and trace a selective history that justifies this. Triumphant, inevitable progress towards this present state of choice is made as a succession of big names tread the path of righteousness. Those who do not fit are dismissed as troublesome heretics, or are ignored.

The outstanding example of Whiggish history of geography is Richard Hartshorne's 1939 treatise on *The Nature of Geography* (see Livingstone, 1992: 8–9 and 304–316). Hartshorne tried to justify his philosophical preference for the discipline's identity – it was to be defined exclusively as the study of areal differentiation – with blatantly Whiggish tactics. His choice of subtitle – *A Critical Survey of Current Thought in the Light of the Past* – was

a dead giveaway, as was the chapter on 'Deviations from the course of historical development'! It was not surprising that Schaeffer's (1953) oft-cited attack on Hartshorne targeted both his philosophy and his history. Schaeffer's paper argued forcefully against 'Exceptionalism in geography' – Hartshorne's belief that the discipline had a special, integrative, mission that set it apart from the other sciences – in favour of a 'scientific' law-seeking approach. Schaefer protested that Hartshorne's view of what geography should be was based on a selective, incorrect reading of past authorities, in particular Kant and Hettner, and produced a counter-argument that was just as Whiggish, supporting his own partial view with an equally selective roll-call of heroic works, including those of von Thünen, Christaller and Lösch. As William Bunge wrote to Hartshorne in 1959, 'history, as conducted in geographic methodological discussions in general, can prove anything and therefore proves nothing' (Livingstone, 1992: 315).

If historical analysis cannot be used to settle philosophical debates it can help us understand why people thought and did certain things. This involves a shift from inwards-looking *textual* accounts of science – an impersonal focus on its logic, methods, data, hypotheses and laws – to outwards-looking, *contextual* accounts that stress subjective factors. Most scientists reject the contention that human inputs make a difference to the knowledge they produce, claiming that the logic and procedures of 'scientific method' neutralize subjectivity and so ensure that the findings of science mirror natural reality; thus, scientific knowledge has a peculiar, *privileged* status. The opposite point of view is that scientific knowledge is always subjective precisely because science is done by people; it is a *social construction*. This perspective is often traced back to Thomas Kuhn's work on scientific revolutions and paradigm shifts, but it was developed in detail by sociologists of science in the 1970s and 1980s (Livingstone, 1992: 12–23). The most extreme versions of this school of thought maintain that it is futile to think that we know anything of a non-human reality; that science is just as subjective as, say, religion or literature; and that any one claim to knowledge is just as good as any other. Livingstone (1992) does not accept this out-and-out *relativism*; if nothing else, the practical achievements of science surely indicate that we must know something about how the world really works. However, he insists that historians of geography should take social factors seriously as part of a history that treats text and context as two sides of the same coin. For example, to understand why Anglo-American geography adopted regional studies with such enthusiasm between the two World Wars, we must look beyond straightforward academic arguments. Powerful figures such as Hartshorne and Carl Sauer (see Livingstone, 1992: 290–303) advocated the turn to particular, synthetic studies of areal differentiation as a strategic move that would distance the discipline from its disreputable environmental determinist past, restore its intellectual and moral purity, and provide a new, distinct identity with which to counter the rise of social sciences such as economics and sociology.

To summarize Livingstone's (1992) core arguments in this first chapter, we might conclude that, firstly, there is no correct way to do geography – it is only part-determined by the state of reality or by philosophical argument. Hence, any history of the discipline that assumes otherwise is wrong, Secondly, contextual, social factors also control the type of geography that is done. These operate at a range of scales, from the individual to society-wide, and change over time and across space. Thirdly, and following from this, '… there can only be a *situated* geography. For geography has meant different things to different people in different places and thus the 'nature' of geography is always negotiated.

The task of geography's historians … is thus to ascertain how and why particular practices and procedures come to be accounted geographically legitimate and hence normative at different moments in time and in different spatial settings' (Livingstone, 1992: 28).

To end the chapter, Livingstone (1992: 30) suggests that we should think of geography as a tradition that evolves over time – an obvious reference to Darwin's theory of species evolution. He describes this as a 'rhetorical flourish', but it is an apt analogy that captures the kind of history he wants to write. Firstly, it negates Whiggish histories, as the trajectory of evolution is contingent, not pre-determined. Secondly, it demands consideration of both internal and external factors, as transmutation of a species or tradition is the product of that entity's (whether organism or idea) interaction with its environment (whether natural or social).

The next eight chapters present case studies intended to illustrate the kind of history that Livingstone advocates. Much of the core subject matter consists of episodes and personalities familiar from the standard textbook histories; for example, the Age of Reconnaissance, Prince Henry the Navigator and Christopher Columbus; the Enlightenment, and James Hutton's discovery of 'deep time' in his *Theory of the Earth*; the scientific voyages of James Cook and Alexander von Humboldt; the formal academic theories of geography's first professors, such as Halford Mackinder, Friedrich Ratzel and William Morris Davis; the struggle for the discipline's scientific identity in the wake of the Schaeffer-Hartshorne clash … all of which I studied as an undergraduate between 1989 and 1992, just before *The Geographical Tradition* was published. The book's most original sections are contributed by Livingstone's own research: on magical geography, neo-Lamarckianism, and – in what is perhaps the book's outstanding chapter – on the applications of environmental determinism that established geography as the practical science of race and empire. The book's final chapter provides a useful summary of what Livingstone considers to be the chief 'conversations', or discourses, that have occurred as the geographical tradition has evolved.

Reception and evaluation

The Geographical Tradition was enthusiastically received: most academics agreed that, despite personal objections to certain of Livingstone's views or omissions, it was by far the best history of geography available. Reviewers highlighted Livingstone's prodigious knowledge of primary and secondary sources, the novelty and power of his contextual stance, the pace and skill of his storytelling, and the elegance of his writing. Denis Cosgrove (1993) described the book as 'a scholarly tour de force'; Audrey Kobayashi (1995) claimed that it was 'absolutely the best choice for those often-resented obligatory courses on the nature of geography', as the 'rave comments' of her students showed; Alan Werrity and Laura Reid (1995) went so far as to suggest that it 'sen[t] one running down the corridor doing cartwheels'! The science journal *Nature* published the only openly hostile review, by historian William McNeill (1993). Under the banner 'Obsessive deconstruction' McNeill attacked what he saw as Livingstone's excessive relativism, arguing that 'the word "truth" is absent from his vocabulary', so that *The Geographical Tradition* was 'intellectual [i.e. ignorant of realities] history with a vengeance'.

Yet Livingstone (1992) was correct to describe the standard histories of geography as boring lists of facts (I recall sitting in one of my first lectures as an undergraduate, diligently writing down that Columbus's three ships were called *Niña*, *Pinta* and *Santa Maria*, just as I had once done at primary school!). *The Geographical Tradition* set new standards of authorship for disciplinary history that was colourful, controversial and relevant. This was a direct consequence of the fresh perspective that Livingstone advanced. His authoritative knowledge of the wider history of science contributed significant new insights that inwards-looking histories of geography, constrained by their concern to keep within artificial disciplinary limits, had missed. For example, standard accounts of the voyages of discovery focused on the new lands and seas added to western Europe's maps; Livingstone saw far more than just new data, arguing that the empirical process of exploration promoted new standards for rational knowledge – prioritizing first-hand experience over established authority – that revolutionized Western science itself (Livingstone, 1992: 32–35 and 59–62). However, it was his motif of the 'contested enterprise' that captured the geographical imagination. The notion of the geographical tradition as something that was subjective, context-dependent and, as a result, flexible, offered a stimulating alternative to self-justificatory histories, often presented in simplistic paradigm terms. Livingstone's (1992) history demonstrated that claims to authoritative knowledge often had more to do with shifting power relationships than with rationality or truth, so that the identity of geography was not fixed, but was up for grabs. This was a message that a large number of geographers starting to explore the possibilities of post-modernism very much wanted to hear. Livingstone (1992) was hesitant to make links to postmodernism in the *The Geographical Tradition*, with no reference to it in the first chapter. However, he did specifically align the postmodern challenge to authority with revisionist historiography in his 1990 essay on 'Geography, tradition and the Scientific Revolution', concluding that:

> ... it is now clear that classical foundationalism [the belief, which underwrites positivist scientific method, that only first-hand experience of some*thing* produces valid knowledge] is in bad shape,

and had better be given up for dead. In this context geographers will have to acknowledge that warranted knowledge is relative to a body of beliefs, not a body of certitudes. Inevitably, this will lead to pluralism in the geographical academy... We now need to realise that non-foundationalist discourses – in the political, the affective, the moral, the artistic, the cultural, the religious, and doubtless a host of other spheres – are as legitimate now as they were in the fifteenth and sixteenth centuries... This has far-reaching consequences. [To address these] is the task that confronts the present generation. (Livingstone, 1992: 370)

Livingstone uses this passage a second time to close his history (Livingstone, 1992: 345–346). With such a call to arms that so obviously tapped into the *Zeitgeist* of human geography (was Livingstone's history just as Whiggish as Hartshorne's?) it is not surprising that *The Geographical Tradition* became an 'instant classic'.

Despite this, *The Geographical Tradition* was controversial. Critics soon identified omissions, with perhaps the most vociferous complaint being that Livingstone's history ignored women. Kobayashi (1995: 194) saw this as 'a flaw so grievous as to be almost fatal'. The feminist objection was matched by post-colonial critics who pointed out that the book focused on European and North American traditions at the expense of others. It is not difficult to understand such criticisms – feminist and post-colonial thinking were an important part of human geography's post-modern turn, which advocated an openness towards multiple (as opposed to privileged) viewpoints as one of its key features. Ironically, Mona Domosh (1991a, 1991b) *supported* her attack on the sexism of David Stoddart's *On Geography* (see below) with reference to Livingstone's previous work advocating alternative histories of geography! There is a second irony too – those

who saw *The Geographical Tradition* as out-of-step with the intellectual cutting edge invoked exactly the kind of presentist critique that Livingstone cautioned against. But what was important in 1992 seems less important now. Livingstone (1992: 30–31) admitted from the start that his case studies were selective so it makes little sense to attack *The Geographical Tradition* for its omissions. If Livingstone and Stoddart wrote histories dominated by privileged, white, western males this was because their subject matter of choice – formal academic geography before the Second World War – was dominated by privileged, white, western males. It is also now recognized that much of the initial work inspired by feminist and post-colonial theories was over-vigorous in its identification of hegemonic conspiracies. The spectre of an all-powerful masculinity was often countered by unsubtle feminism that, despite its claim to respect pluralities, was just as essentializing. Subsequent work has tended to be more nuanced in its discussion of identity – see, for example, McEwan (1998) – in part because it displays greater sensitivity to the particularities of place. This need for careful historical-geographic study of the history of geography is evident in *The Geographical Tradition*, and has been developed explicitly in Livingstone's subsequent work.

It is perhaps more pertinent to ask if *The Geographical Tradition* matches Livingstone's own objectives. Several of his case studies unequivocally demonstrate the need for contextual histories. For instance, we now see environmental determinism as both blatantly racist and scientifically absurd – but, because it set geography up as *the* science of empire, it endured as a core component of the geographical tradition for over 50 years (Livingstone, 1992: Chapter 7). This example clearly shows that the success of science cannot be read off as a simple function of the extent to which it matches up to real-world truth. Elsewhere, however, Livingstone's history is less convincing. His account of developments after *c.* 1970 is disappointing. Temporal proximity cannot satisfactorily excuse what is little more than a list of big names and big ideas – just the type of history Livingstone wanted to discard. For what is supposed to be a contextual history, it is surprising that he makes almost no attempt to examine the relationship between postmodern thought and the specific economic, political and cultural phenomena indicative of postmodernity outside the ivory towers (see Chapter 15). Driver (2004: 231), was right to point out that *The Geographical Tradition* is 'a very book-ish book', in that it tends to prioritize the details of particular ideas. This is evident in Livingstone's discussion of evolution. If positivism was important to twentieth-century geography because of its style, not its substance (Livingstone, 1992: 321–322; see also Taylor, 1976), should it not also be the spirit, not the detail, which accounts for the influence of evolutionary doctrines on nineteenth-century geography? Yet Livingstone concentrates on the differences between Lamarckian and Darwinian versions of evolution. This narrow focus on ideas is at odds with studies by sociologists of science, introduced in *The Geographical Tradition*'s first chapter, which shows that it is important to follow what scientists *do* if we are to understand how and why knowledge arises (see Demeritt, 1996, for a review of this work).

It is true that *The Geographical Tradition* demonstrated a new way to write the history of geography – but it was not the first text to do so. Papers such as Taylor's (1976) interpretation of the quantification debate as a social power struggle, in which the revolutionaries used mathematics as a weapon against the established regime, or Peet's (1985) exposé of environmental determinism as the ideological tool of imperial capitalism, had previously emphasized the importance of subjective factors. The first single-author book by a

geographer to take contextual history seriously was David Stoddart's *On Geography and its History*, published in 1986. Stoddart's (1986) first chapter presents a manifesto for contextual histories of the discipline that anticipates much of *The Geographical Tradition*'s first chapter. Livingstone's enthusiastic review appeared in *Progress in Human Geography* in 1987. So, if the two books are similar, why is it that *The Geographical Tradition*, not *On Geography*, became established as the breakthrough text? Both are impressive works of scholarship, although written in very different styles. But Livingstone's history differed from Stoddart's in several key aspects: he made more use of work from the history and sociology of science; he was more flexible in his attitude towards what geography was, or should, be; he was far more sceptical about the power of empirical scientific method; and he was more sensitive to matters of discourse and representation. This did not make *The Geographical Tradition* a better book, but it did make for a better match with the wants and expectations of those attracted by the possibilities of post-modernism. As Livingstone's recent research shows (Livingstone, 2005), the context in which a book is read can be just as important as what it says. Written by a human geographer with strong affinities for the humanities (Stoddart is a physical geographer), and published six years after Stoddart's book in 1992, by which time the 'cultural turn' was in full swing, the impact of *The Geographical Tradition* must itself be understood in both textual and contextual terms. It was very much a case of the right message in the right place at the right time.

Conclusion

Because of *The Geographical Tradition*'s timely appearance amidst a sea of persuasive, complementary ideas (see Driver, 1995), it is important not to claim too much for a single book. But it is clear that *The Geographical Tradition* inspired geographers to rethink the foundations of their discipline. Livingstone's (1992) notions of the 'geographical tradition' and the 'contested enterprise' provided powerful metaphors that soon became part of the discipline's everyday language, uniting geographers with diverse outlooks – in debate, if not agreement (Mayhew *et al.*, 2004: 229). Undergraduate courses on the history and philosophy of geography had to be re-written. The history of geography, often thought of as the poor relation of 'proper', contemporary research, received a major boost. By placing them firmly in the orbit of postmodernism *The Geographical Tradition* helped to make historical studies of knowledge both fashionable and relevant (but see Barnett, 1995, for a very different view of history's relevance). If the first of the new batch of historical case studies took a predictable direction under the influence of feminism and/or post-colonialism – Barnett (1995: 418) commented that 'it would be quite disastrous for the theoretically inclined human geographer if their discipline did *not* have a dubious imperialist past' – more recent work has opened up different times and places. Much of this new work foregrounds the actions of particular personalities in particular localities, and so corrects the tendency of past case studies, including some of those in *The Geographical Tradition*, to isolate the intellectual. It is now common for geography's historians to promote the principle – pleasingly geographical, but firmly non-rational – that the *who* and the *where* of scientific

enquiry make a difference to the content of the knowledge produced. Livingstone's contribution is important here (Livingstone, 2003, 2005), but perhaps the outstanding example – because, unusually, it examines near-contemporary history – is Trevor Barnes's work which sets out a 'post-prefixed' interpretation of economic geography (e.g. Barnes, 1996, 2001; see Kelly, Chapter 23 this volume).

It is ironic that, by stimulating cross-disciplinary revival of historical and philosophical debate, *The Geographical Tradition* undermined its core premise – that it makes sense to study a single body of beliefs and actions that constitute *the* geographical tradition. Livingstone (1992: 420) dislikes what he called 'the postmodern pluralization imperative' but his argument that *the* tradition exists simply because it is possible to tell its story, however varied that story might be, looks far from convincing 15 years on. Much of what geography had, or still has, in common comes from the contingent co-location of its practitioners in Departments of Geography. Persistent attempts to start conversations across assorted divides (e.g. Harrison *et al.*, 2004) perhaps indicate a collective desire for disciplinary identity founded on necessity, but often do little more than emphasize the multitude of traditions in which geographers now participate. If we follow through Livingstone's analogy of tradition-as-species, this pluralism should not surprise us (or him), for as species evolve differences accumulate, so that, at some point and in some place, new species distinct from, and often incapable of fertile interaction with, their ancestors appear.

Secondary sources and references

Barnes, T.J. (1996) *Logics of Dislocation*. New York: Guilford Press.

Barnes, T.J. (2001) 'Lives lived, and lives told: biographies of geography's quantitative revolution', *Environment and Planning D: Society and Space* 19 (4): 409–429.

Barnett, C. (1995) 'Awakening the dead: who needs the history of geography?' *Transactions of the Institute of British Geographers* NS 20 (4): 417–419.

Cosgrove, D. (1993) 'Book review of *The Geographical Tradition*', *Progress in Human Geography* 17 (4): 583–585.

Demeritt, D. (1996) 'Social theory and the reconstruction of science and geography', *Transactions of the Institute of British Geographers* NS 21 (3): 484–503.

Domosh, M. (1991a) 'Toward a feminist historiography of geography', *Transactions of the Institute of British Geographers* NS 16 (1): 95–104.

Domosh, M. (1991b) 'Beyond the frontiers of geographical knowledge', *Transactions of the Institute of British Geographers* NS 16 (4): 488–490.

Driver, F. (1995) 'Geographical traditions: rethinking the history of geography', *Transactions of the Institute of British Geographers* NS 20 (4): 403–404.

Driver, F. (2004) 'Commentary two (Classics in Human Geography Revisited – the Geographical Tradition)', *Progress in Human Geography* 28 (2): 229–233.

Harrison, S., Massey, D., Richards, K.S., Magilligan, F.J., Thrift, N. and Bender, B. (2004) 'Thinking across the divide; perspectives on the conversations between physical and human geography', *Area* 36: 435–442.

Kobayashi, A. (1995) 'Book review of *The Geographical Tradition*', *Annals of the Association of American Geographers* 85 (1): 191–194.

Livingstone, D.N. (1990) 'Geography, tradition and the Scientific Revolution: an interpretative essay', *Transactions of the Institute of British Geographers* 15 (3): 359–373.

Livingstone, D.N. (1992) *The Geographical Tradition: Episodes in the History of a Contested Enterprise*. Oxford: Blackwell.

Livingstone, D.N. (1995) 'Geographical traditions', *Transactions of the Institute of British Geographers* NS 20 (4): 420–422.

Livingstone, D.N. (2003) *Putting Science in its Place*. Chicago: University of Chicago Press.

Livingstone, D.N. (2005) 'Science, text and space: thoughts on the geography of reading', *Transactions of the Institute of British Geographers* NS 23 (4): 420–422.

McEwan, C. (1998) 'Cutting power lines within the palace? Countering paternity and euro-centrism in the "geographical tradition"', *Transactions of the Institute of British Geographers* NS 23 (3): 371–384.

McNeill, W. (1993) 'Obsessive deconstruction. Book review of *The Geographical Tradition*', *Nature* 362: 218 (18 March 1993).

Mayhew, R., Driver, J.F. and Livingstone, D.N. (2004) 'Classics in human geography revisited: Livingstone, D.N., 1992, *The geographical tradition: episodes in the history of a contested enterprise*', *Progress in Human Geography* 28 (2): 227–234.

Peet, R. (1985) 'The social origins of environmental determinism', *Annals of the Association of American Geographers* 75 (3): 309–333.

Stoddart, D.R. (1986) *On Geography and its History*. Oxford: Blackwell.

Taylor, P.J. (1976) 'An interpretation of the quantification debate in British geography', *Transactions of the Institute of British Geographers* NS 1 (2): 129–142.

Werrity, A. and Reid, L. (1995) 'Debating the geographical tradition', *Scottish Geographical Magazine* 111 (3): 196–197.

19 FEMINISM AND GEOGRAPHY (1993): GILLIAN ROSE

Robyn Longhurst

In exploring the masculinism of geography at some length, this is not a book about the geography of gender but about the gender of geography. (Rose, 1993: 4–5)

Introduction

Gillian Rose wrote *Feminism in Geography: the limits of geographical knowledge* early in her career when she was Lecturer in Geography at Queen Mary and Westfield College, University of London. Soon after, Rose was appointed Lecturer in Geography at the University of Edinburgh, moving later to a Professorship of Geography at the Open University. Prior to the publication of *Feminism and Geography*, feminist geographers had focused attention mainly on empirical studies of women. The 1984 landmark *Geography and Gender* by the Women and Geography Study Group (see Hanson, Chapter 11 this volume) contained several chapters that cover a range of feminist theoretical and methodological approaches, but focused mainly on the effect of urban spatial structures on women, women's employment, women's access to facilities, and women and development. *Geography and Gender* thus laid an important foundation for feminist geography but its primary aim was to examine 'geographies of gender'. Rose's primary aim in *Feminism and Geography* was to examine the 'gendering of geography'. At the beginning of the 1990s geography was

severely in need of such an in-depth and sustained feminist examination.

Reception and evaluation

When *Feminism and Geography* arrived in the mail (it was not available in New Zealand bookstores so I had ordered it) I was writing a feminist-informed doctoral thesis on pregnant bodies in public spaces. I was trying to think through how bodies (which are commonly associated with Woman, irrationality, and the private sphere) had been excluded in geography but I wasn't quite sure which bodies had been excluded, in which geographies, and what purpose this served. I had been looking forward to the arrival of Rose's book hoping it might cast some light on these issues. I was not disappointed. I devoured the 200 page text over a weekend (it was an intense read because, as Karen Morin, 1995, notes 'a lot is packed into this small volume'). The following Monday I recall contacting my doctoral supervisor to report that Rose's book had prompted a breakthrough in my research.

Needless to say *Feminism and Geography* didn't just have an impact on me. At the time of its publication, the listserv 'Feminism in Geography' (GEOGFEM@LSV.UKY. EDU) had not long been established and the book quickly became the subject of much discussion most of which centred upon unpacking the ideas contained within the text. There is a reasonably heavy reliance throughout the book on academic prose which makes the work challenging, especially for new or emerging scholars. In a review of *Feminism and Geography*, Briavel Holcomb (1995: 264) says the book is written in 'the arcane language of critical social theory'. The listserv discussion was invaluable in that it provided an arena for many of us to share our ideas and opinions on the many and complex arguments contained in the book.

However, not only was *Feminism and Geography* discussed informally on listserves such as 'Feminism in Geography' but it was also reviewed extensively in geography journals such as *Annals of the Association of American Geographers* (Falconer Al-Hindi, 1996), *The Geographical Journal* (Burgess, 1994), *Transactions of the Institute of British Geographers* (Kofman, 1994) and *Geographical Review* (Holcomb, 1995). The book's influence also extended beyond geography. It was reviewed in feminist studies journals such as *Feminist Review* (Bondi, 1995) and *Signs* (Hayden) and in journals in other disciplines such as *The Journal of American History* (Norwood, 1994) and *Postmodern Culture* (Morin, 1995).

Rose (1993: 1) begins *Feminism and Geography* by explaining 'The academic discipline of geography has historically been dominated by men, perhaps more so than any other science'. Men have been over-represented in professional geography organizations, in geography departments and in academic publishing. Men's interests, therefore, have structured what counts as legitimate geographical knowledge. Rose makes the point that women's exclusion in geography is not just about the themes of research, nor even about the concepts feminists employ to organize those themes, but that there is something in the very claim to knowing in geography which tends to exclude women as producers of geographical knowledge.

Rose (1993: 15) says she 'desperately wanted to be able to join in, to be part of debates among knowledgeable men' and was 'seduced' by the academy but occupied an ambiguous position. Rose was empowered by her whiteness but at the same time marginalized as a woman. When I first enrolled as a young, white, working-class, female university student in 1980 I too felt enormously attracted to the academy and in particular to the discipline of geography but also excluded from it. I felt excluded from discussions and activities in lectures, tutorials, and on fieldtrips. I also felt excluded when I chose to study topics that were deemed to be inappropriate for geographical study, such as in 1985 when I wanted to write a Master's thesis on sexual violence against women. Many years later as a Lecturer I revelled in reading, teaching and researching geography but still felt like an impostor in the discipline, not just on account of the topics I chose to study (I was once told by a geography journal editor that pregnant bodies were an inappropriate subject for geography) but at a much deeper level. Rose's text spoke to me. I too felt positioned uneasily in the discipline. *Feminism and Geography* provided me with a framework within which to understand further my sense of being both an insider and an outsider in geography.

Feminism and Geography challenges geographers to think critically about the way geographers produce particular kinds of knowledge. Rose (1993) argues there is a specific notion of knowing, and of knowledge, as masculine, exhaustive, rational, and associated with the mind rather than the body (as though the two can be separated) which marginalizes women in the production of geographical knowledge.

In order to mount this argument about the intersection of power, embodied subjectivity and knowledge, Rose (1993) relies heavily on the intellectual cross-fertilization of feminist poststructuralist, psychoanalytical and geographical theory. Rose was highly influenced by Doreen Massey's (1984: 129) pioneering work on the *Spatial Divisions of Labour* (see Phelps, Chapter 10 this volume) which she says encouraged 'studies of the intricate geography of gender and class accessed through local studies'. But Rose also draws on the feminist theorizing of Rosi Braidotti, Marilyn Frye, Diana Fuss, Moira Gatens, Donna Haraway, Luce Irigaray, Michéle Le Doeuff, Elspeth Probyn, and Iris Marion Young. She also draws on the work of queer theorists such as Teresa de Lauretis and Eve Sedgwick. Contributions from black and postcolonial feminists such P. Hill Collins, bell hooks and Gatari C. Spivak are also used to further the argument that geography is a 'masculinist' and white discipline.

Rose adopts the term 'masculinist' from Michéle Le Doeuff (1991: 42, cited in Rose, 1993: 4) who defines it as 'work which, while claiming to be exhaustive, forgets about women's existence and concerns itself only with the position of men'. Rose describes geography as 'masculinist' because it has produced grand theories that claim to speak for everyone but in actual fact speak only for heterosexual, white, bourgeois men. The substance of Rose's argument is that there are at least two different kinds of masculinity at work in the discipline. She refers to the first as 'social-scientific masculinity'. This kind of masculinity 'represses all reference to its Other in order to claim total knowledge' (1993: 10–11). Rose critiques the 'social-scientific masculinity' in the time-geography of Swedish geographer Törsten Hägerstrand (see Lentrop, Chapter 1 this volume). She explains that while time-geography was useful, in that some of the earliest feminist geographers adopted it in order to 'recover the everyday and the ordinary' (Rose, 1993: 22) that are often seen to typify women's lives, it was not able to account for the kind

of embodied subjectivity produced by the routine work of mothering and domesticity that occurs mainly in the home. Hence, 'women and their feelings somehow got lost' (Rose, 1993: 27) in time-geography.

Rose refers to the second kind of masculinity in geography as 'aesthetic masculinity'. This kind of masculinity 'establishes its power through claiming a heightened sensitivity to human experience … [it] admits the existence of its other in order to establish a profundity of which it alone has the power to speak' (Rose, 1993: 11). 'Aesthetic masculinity' can be found in humanistic geography which establishes place as key concept, partly as a response against positivism which became so popular in the discipline in the 1960s. 'Man' was to be put back into the centre of things (Rose, 1993: 43). Humanistic geographers, however, tend to feminize place which is conceptualized as the idealized Woman. They seek to know place exhaustively, while at the same time asserting its mysterious unknowability.

Rose's critique of masculinism, however, does not stop with time-geography and humanism. She also critiques a strand of radical feminism by examining the dualism between Nature (feminine) and Culture (masculine). Geographers often display contradictory impulses towards Nature (feminine). They fear Nature and want to dominate it but also desire it, revelling in its magnificence and beauty. Rather than problematize this dualism, many radical feminists simply invert it. Rose also turns her attention to cultural geography, in particular to the concept of landscape. Cultural geographers, she argues, fail to address their own pleasure in looking at landscape (see Gilbert, Chapter 12 this volume).

Rose understands feminist geography to be both complicit with and resistant to masculinism. In some ways feminist geographers have challenged the masculinism inherent in the discipline (for example, by considering production *and* reproduction) but in other ways they have been complicit with masculinism (for example, by invoking a conception of Woman that loses sight of the diversity amongst women). This critical mobility that positions feminists both inside and outside of geography, as both complicit and resistant to its masculinism, offers strategic possibilities for 'imagining something beyond the discipline's hegemonic imagination' (Rose, 1993: 117).

Impacts and effects

Feminism and Geography was hugely important in propelling ideas in certain directions in the discipline of geography. First, it encouraged geographers, including *feminist* geographers, to reflect critically on their/our own role as producers of knowledge. Rose (1993) acknowledges that feminism itself is caught in existing masculinist discourses and that there is no space of purity from which Woman or women can speak. She acknowledges the challenge that feminism faces in needing to build an identity for Woman, while at the same time recognizing the diversity amongst women in relation to race, class, and sexuality and calls for a 'strategic' or 'critical' mobility (Rose, 1993: 13) as a form of feminist resistance. This idea has contributed richly to arguments about researcher reflexivity and postionality that began to emerge in the discipline in the early 1990s (e.g. see McDowell, 1992 and a 'Special Issue of Women in the Field' in *Professional Geographer*, 1994).

A second contribution Rose made to both geography and feminist studies was to further the critique of dualistic thinking. Other feminists, including feminist geographers such as Liz Bondi (1992) and Dina Vaiou (1992) had, prior to the publication of Rose's (1993) book, begun to illustrate how geographical discourse focused on the distinctions between culture/nature, public/private, production/reproduction, western/oriental, work/home, and state/ family. They also made the argument that these dualisms were gendered but Rose was the first to offer a book-length, sustained engagement with this notion.

The third contribution *Feminism and Geography* made was to prompt work on 'embodied geographies'. Rose (1993: 7) claims 'Masculinist rationality is a form of knowledge which assumes a knower who believes he can separate himself from his body, emotions, values, past experiences and so on'. This allows for him to consider his thoughts (his mind) to be autonomous, transcendent and objective; mess and matter-free so to speak. She points out that 'the assumption of an objectivity untainted by any particular social position', or any particular body, allows masculinist rationality to 'claim itself as universal' (Rose, 1993: 7). These ideas helped give rise to a new area of study – 'embodied geographies' (e.g. Duncan, 1996 and Teather, 1999). Lise Nelson and Joni Seager (2005: 2) claim 'The body is the touchstone of feminist theory' but up until the publication of *Feminism and Geography* feminist geographers had been more influenced by liberal and socialist, rather than radical, versions of feminism in which 'the body' was not centre-stage.

A fourth impact the book had was to open up the discipline of geography to the possibility of using feminist psychoanalysis to understand better the relationship between the social, spatial and psychic. In turn, Rose opened up the discipline of feminist psychoanalysis to the possibility of focusing on issues of spatiality. Rose claims that feminist psychoanalytic commentaries offer an eloquent critique of geography's white, heterosexual, masculine gaze, a gaze torn between pleasure and its repression (Rose, 1993: 103). In 1993 only a handful of geographers were engaging with psychoanalysis, perhaps most notably Steve Pile, who in 1996 published *The Body and the City*. It was not until the latter half of the 1990s that feminist geographers and others really began to engage with psychoanalytic approaches to bodies in an attempt to understand more about the psychological acquisition of gendered, sexed and racialized identities and relationships with others (e.g. Nast, 2000).

A fifth contribution made by *Feminism and Geography* is Rose's reimagining of a geography that is not based on masculinism or a dualist system that dominates the Self and marginalizes the Other. Rose (1993) asks how might it be possible to challenge the discipline so that hegemony of Man/Woman gives rise to geographies of difference? To answer this question she puts forward the notion of 'A Politics of Paradoxical Space' that acknowledges the power of hegemonic discourse but also insists on the possibility of resistance (Rose, 1993: 155). Feminism needs to occupy both the centre and the margin – to be mobile, multiple, and contradictory – so as to 'threaten the polarities which structure the dominant geographical imagination' (Rose, 1993: 155). This concept, which entails a radical rethinking of place, space and gender, has opened up possibilities for new ways of thinking about people–place relationships (see Bondi and Davidson, 2005, 20–25 on how feminist geographers have used the concept of 'paradoxical space').

Finally, Rose's (1993) text changed the way that geographers and others in disciplines such as landscape architecture, urban planning, feminist studies and cultural studies engage with the visual. In Chapter Five – 'Looking at Landscape: The Uneasy Pleasures of Power' – Rose opens up for question the politics of looking, and the conflation of seeing and knowing. This strikes to the heart of the discipline. This is an area that Rose herself has developed in theoretical and empirical research over the past decade (e.g. see Rose, 2003). She has also written about methodologies for interpreting visual materials (Rose, 2001). Looking, seeing, and knowing (the package deal offered by most geography fieldtrips) have long been mainstays of spatial disciplines such as geography.

Despite these contributions the book has not been without it critics. Holcomb (1995: 264) suggests that Rose's 'sweeping condemnation of most male-produced geography suggests the very tendency to essentialize of which she accuses men'. Holcomb explains that she herself became a geographer because she 'loved fieldwork' and so was dismayed to read that this 'initiation ritual [of] tough heroism establishes fieldwork as a particular kind of masculine endeavour' (Rose, 1993, cited in Holcomb, 1995: 264–265).

Eleonore Kofman (1994: 496) notes that 'the repeated and insistent attacks on the textual strategies of male cultural geographers left [her] reeling'. Kofman (1994: 497) claims she is 'not at all convinced that oscillating between difference and unity … will undermine masculinism'. Kofman also critiques the book on the grounds that it lacks historical depth in the recounting of the masculine narrative of the discipline and that 'there are a number of epistemological issues which have influenced the positions adopted in the book but which have not been made explicit' (Kofman, 1994: 497).

Vera Norwood (1994: 834) reiterates some of Kofman's points. Norwood argues there is insufficient historical analysis of key men and documents in *Feminism and Geography* to 'convince even a sympathetic reader' of Rose's case. She says Rose (1993) criticizes men in the discipline of creating overly abstracted and disembodied figures of a generic Woman but she 'falls in the same practice in her descriptions of male colleagues' (ibid.). Norwood (1994: 833) argues '*Feminism and Geography* is a very ambitious book that often falls short of its project'. Some reviewers also criticized the 'production' of the book. Kofman (1994) notes the book does not contain an alphabetically ordered bibliography, which means the reader is endlessly searching the notes for references. Morin (1995) points out that 'Occasional misspellings don't help matters, and the book's three illustrations are merely adequately reproduced'.

Conclusion

Like any book, this one undoubtedly has its pros and cons, its supporters and detractors. It has been criticized on some fronts but overall I think it is fair to say the book has received overwhelmingly positive reviews. Jacquelin Burgess (1994: 226), for example, describes the book as 'subtle and sophisticated'. Liz Bondi (1995: 133) says 'Rose's critique of geographical knowledge is powerful and far-reaching'. Since its publication in 1993 *Feminism and Geography* has undoubtedly proven itself to be a key text. It has been read widely and debated vigorously. Rose succeeded in producing a book that is indispensable for any feminist geographer or other scholar attempting to come to grips with feminist and/or geographical thought. Linda McDowell, on the back cover of *Feminism and Geography*, comments: '[Rose's] book will become essential reading for everyone interested in philosophical and methodological issues in geography'. McDowell was correct; the book has become essential reading. Of course there are still some who would disagree and consider Rose's (1993) insights to be misinformed, unimportant and/or peripheral to the history of geography. But maybe this is the real testimony to the book's success; in some ways it *hasn't* slotted in easily to the geographical canon because it still poses a radical challenge to much of that canon. Fourteen years after publication the book itself still occupies something of paradoxical space. Whilst in some ways it has become a key text in the discipline, in other ways it remains 'illegitimate' and unlikely to be read by those who would benefit most from its insights.

Secondary sources and references

Bondi, L. (1992) 'Gender and dichotomy', *Progress in Human Geography* 16 (1): 98–104.

Bondi, L. (1995) 'Review of *Feminism and Geography* by G. Rose', *Feminist Review* 51: 133–135.

Bondi, L. and Davidson, J. (2005) 'Situating Gender', in L. Nelson and J. Seager (eds) *A Companion to Feminist Geography*. Malden, MA: Blackwell, pp.15–31.

Burgess, J. (1994) 'Review of *Feminism and Geography* by G. Rose', *The Geographical Journal* 160 (2): 225–226.

Duncan, N. (1996) (ed.) *BodySpace*. London: Routledge.

Falconer Al-Hindi, K. (1996) 'Review of *Feminism and Geography* by G. Rose', *Annals of the Association of American Geographers* 86 (3): 610–611.

Hayden, D. (1997) 'Review of *Space, Place and Gender* by D.B. Massey and *Feminism and Geography* by G. Rose', *Signs* 22 (2): 456–458.

Holcomb, B. (1995) 'Review of *Feminism and Geography* by G. Rose and *Gender, Planning and the Policy Process* by J. Little', *Geographical Review* 85 (2): 262–265.

Kofman, E. (1994) 'Review of *Feminism and Geography* by G. Rose', *Transactions of the Institute of British Geographers* 19 (4): 496–497.

Le Doeuff, M. (1991) *Hipparchia's Choice: an Essay Concerning Women, Philosophy, etc*. Oxford: Blackwell.

McDowell, L. (1992) 'Doing gender: feminism, feminists and research methods in human geography', *Transactions of the Institute of British Geographers* 17: 399–416.

Massey, D. (1984) *Spatial Divisions of Labour.* Basingstoke: Macmillan.

Morin, K. (1995) 'Review essay: the gender of geography'. Review of *Feminism and Geography* by G. Rose, *Postmodern Culture* 5: 2.

Nast, H.J. (2000) 'Mapping the "unconscious": racism and the Oedipal family', *Annals of the Association of American Geographers* 90 (2): 215–255.

Nelson, L. and Seager, J. (2005) 'Introduction', in L. Nelson and J. Seager (eds) *A Companion to Feminist Geography.* Malden, MA: Blackwell, pp.1–11.

Norwood, V. (1994) 'Review of *Feminism and Geography* by G. Rose', *The Journal of American History* 18 (2): 833–834.

Pile, S. (1996) *The Body and the City: Psychoanalysis, Space and Subjectivity.* London: Routledge.

Professional Geographer (1994) Special Issue on Women in the Field 46 (1).

Rose, G. (1993) *Feminism and Geography.* Cambridge: Polity Press.

Rose, G. (2001) *Visual Methodologies: An Introduction to the Interpretation of Visual Materials.* London: Sage.

Rose, G. (2003) 'Just how, exactly, is geography visual?' *Antipode* 35: 212–221.

Teather, E. (1999) (ed.) *Embodied Geographies: Spaces, Bodies and Rites of Passage.* London: Routledge.

Vaiou, D. (1992) 'Gender divisions in urban space: Beyond the rigidity of dualist classifications', *Antipode* 24 (2): 247–262.

Women and Geography Study Group of the IBG (1984) *Geography and Gender: An Introduction to Feminist Geography.* London: Hutchinson.

20 GEOGRAPHICAL IMAGINATIONS (1994): DEREK GREGORY

John Pickles

All I seek to do is make a series of incisions into the conventional historiography of geography and show that its strategic episodes can be made to speak to many other histories. (Gregory, 1994: 14)

Introduction

From his first book, *Ideology, Science and Human Geography* (1979), Gregory has been one of the most erudite voices arguing for the importance of critical theory in geography, shaping its form and interpreting its wider possibilities for geographers and non-geographers alike. Through *Regional Transformation and Industrial Revolution* (1982) and *Social Relations and Spatial Structures* (Gregory and Urry, 1985) to *Geographical Imaginations* (1994), and subsequently through *The Colonial Present* (2004), and, with Allan Pred, *Violent Geographies* (2006), Gregory has charted a complex and broad geographical project of critical theory. This is a project of re-working geographical theory to sustain a conversation about historical materialism and human agency, and in a way that resists the artificiality of disciplinary boundaries and institutions. For Gregory, disciplines are enabling institutions, but he insists they must not be binding limits. Instead, he proposes critical human geography as part of a trans-disciplinary (even post-disciplinary) project integrating geographical, social, and cultural theory to understand the production of everyday bodies, places, and spaces.

Founded on a theory of history and geography that rejects disciplinary hagiography and grand theory, *Geographical Imaginations* seeks to model a politics of sense-making without universals or absolutes, one that questions grand modernist narratives in favor of more grounded and contextual cultural theory. *Geographical Imaginations* was published in 1994 in the wake of Ed Soja's *Postmodern Geographies* (1989, see Minca, Chapter 16 this volume), David Harvey's *The Condition of Postmodernity* (1989, see Woodward and Jones, Chapter 15 this volume), the English translation of Henri Lefebvre's *The Production of Space* (1991), and Susan Buck-Morss's rendering of Walter Benjamin's Arcades Project in *Dialectics of Seeing* (1989), all texts which problematize the modernist project and with which *Geographical Imaginations* engages in turn, exploring the connections between human geography and critical social theory, and teasing out from them the political value of theoretical work and the inherently geopolitical role of theory.

Writing against 'the deadening proclamations about the "nature" or "spirit and purpose" of geography' Gregory (1994: 78) turns instead to political economy, social theory, and cultural studies, which he defines 'as a series of overlapping, contending and contradictory discourses that seek, in various ways and for various purposes, to reflect explicitly and more or less systematically on the constitution of social life, to make social practices intelligible and to intervene in their conduct and consequences.' As David Harvey (1995: 161) has pointed out: 'It is plainly his intention to take what he considers the very best of geographical work and treat it on a par with some of the very best writings from philosophy, sociology, anthropology, and the like and to illustrate the major contributions that geographers have made, and are making, and can make to social and literary theory.' Both Harvey (1995) and Katz (1995) have questioned the wisdom in Gregory's work of eliding the geographical imaginations at work in physical geography and the natural sciences, and this is partly the case in *Geographical Imaginations*, but in terms of its engagements with the humanities and social sciences *Geographical Imaginations* remains among the most well developed and articulated texts of critical geographical theory. What began as an attempt to explain geographical ideas and history for geographers ended up as a text about geographical ideas and imaginations well beyond the limits of disciplinary boundaries.

Background: the book and its author

As with any text, it is useful to read *Geographical Imaginations* as an elaboration of the author's earlier works, especially *Ideology, Science and Human Geography*, in which he interrogated three central theoretical traditions in geography: positivism, Marxism, and humanism. In this earlier book he developed a sustained critical theory of the history and nature of positivism (from Auguste Comte to the logical positivist and logical empiricists and their goal of unified science to the emergence, promise, and dominance of spatial science in Geography), humanist approaches to Geography (including phenomenology, hermeneutics, and interpretative approaches more generally), and critical emancipatory geographies (such as Marxism).

Three aspects of this earlier work remain especially important in *Geographical Imaginations*. First, Gregory draws on Frankfurt School critical theory and particularly Jürgen

Habermas's *Theory and Practice* (1988) and related writings to develop an epistemological critique of essentialism and functionalism in spatial science, and to demonstrate the always interested and political commitments of theory. Second, in *Regional Transformation* he takes up the challenge of Marxist historian E.P. Thompson to write historical geography and political economy from the perspective of the everyday struggles of people in particular places, not as products of structures taking shape behind their backs but as active agents shaping their worlds and making their own histories, albeit not under conditions of their own choosing:

> Thompson regards historical eventuation as an existential struggle whose recovery requires a recognition of 'the crucial ambivalence of our human presence in our own history, part-subjects, part-objects, the voluntary agents of our own involuntary determinations.' For, if history is 'unmastered human practice,' and if its subjects are 'ever-baffled and ever-resurgent human agents' whose effectivity 'will not be set free from ulterior determinate pressures nor escape determinate limits,' there is nevertheless a space for the insistent return of conscious, knowledgeable *agency*. (Gregory, 1982: 9–10)

Third, in *Social Relations and Spatial Structures* (edited with John Urry, 1985) and subsequently in *Human Geography: Society, Space and Social Science* (edited with Ron Martin and Graham Smith, 1994), he focused more directly on the role of 'human agency in human geography' in ways that highlight the connections between space and society, structure and agency, and economy and culture. His engagements with historical geography, political economy, and cultural politics illustrate this commitment to conceptual border crossing that sustains the political power of theoretical work – a power that 'lies in some part in its power to interrupt, displace, and call into question the taken-for-granted of the world and our place in it' (Gregory, 1995: 177).

Geographical Imaginations ranges across a broad intellectual landscape. Throughout, Gregory reads texts as opening possibilities for dialogue among different traditions. As he writes:

> It has to be possible – and it is an important part of my project, in these essays and elsewhere – to rise above the cynical disparagement of theoretical work, to interrogate its *other*, creative, imaginative and productively political values. (Gregory, 1994: 49)

The result is a rich and heady wine; a cultural politics of spatial practices that ranges across the mapping impulse, spatial analysis, contemporary urbanisms, social action, and the production of space.

Structure and arguments

Geographical Imaginations is organized in three parts, each with an introductory essay and two chapters. Part One, '*Strange Lessons in Deep Space*', was written last and can be seen as a charting of the ground for his subsequent writing on Edward Said and the 'Middle East.' It focuses on the ways in which space and representation have been thought and acted upon. He begins with the visual regime that 'rendered' the world as a representation – i.e. as a picture, an exhibition, and as a mere object-form. Gregory shows how this visual regime worked to shape the colonial gaze and with it the universalist claims of European science. This was the 'god-trick' that produced nature and its local inhabitants as merely resources for settlement and exploitation. It was also the founding moment for the creation of a view of knowledge that assumed a separation between subject–object, and it gave rise to a deep epistemological break in geographical thinking; a cartographic anxiety that parallels what Richard

Bernstein (1983) had characterized as the Cartesian Anxiety. Part One of *Geographical Imaginations* clarifies the hold this visual regime has on contemporary social and spatial thought and begins the process of reworking these naturalized notions of space and representation, 'remapping spaces of power-knowledge' (Gregory, 1994: 33) to develop a more thorough-going conjunctural cultural analysis of place, space, and landscape. He maps out the ways in which modern spatial science has been 'socialized' by its engagement with political economy, social theory, and cultural studies, and correspondingly how questions of place, space, and landscape have been taken up in other humanities and social sciences, particularly through philosophy, feminism, post-structuralism, and post-colonialism.

Part Two, '*Capital Cities*', turns more directly to the connections between space and representation, politics and poetics, and the role of the city in shaping contemporary theory and politics. He reads 'capital cities' through the variously modernist and postmodernist urban mappings of David Harvey, Walter Benjamin, Allan Pred, and Ed Soja, drawing from each the ways in which they foreground 'capital cities' as generative sites of cultural capital, but also as privileged sites of cultural innovation and theory production (Gregory, 1994: 213). This is a cosmopolitanism with which he is increasingly uneasy and one that leads him closer towards Said. As Gregory writes on the first page of the book, this in itself is an act of migration, a re-centering, and a displacement, first through his own move from Cambridge to Vancouver (where he more directly recognized the challenges and possibilities of post-colonial, multi-cultural and gendered politics and theory) and second through the ways in which he seeks to open conversations among a broad range of contextual and anti-essentialist social and cultural theories. While commentators have expressed surprise that post-colonial, multi-cultural and feminist movements had not apparently reached Cambridge (Barnett, 1995; Harvey, 1995; Swyngedouw, 1995), Gregory (1995) prefers to describe his move to Canada in 1989 as an important symbolic and rhetorical border-crossing from a Cambridge and Britain of discipline and metropolitanism to a Vancouver and Canada in which intense struggles over social and political identity were being fought in and beyond the university in the postcolony (see Sparke, 2006).

Part Three, 'Between Two Continents' investigates further the 'uncomfortable' space between historical materialism and postmodernism. Gregory reads Harvey's The Condition of Postmodernity and those who, in responding to its arguments, have struggled with the ways in which it addresses the tension between theory's desire for unity and a fear of 'insufficiency, absence, fragmentation.' He draws on Walter Benjamin's critical interventions in Marxism and cultural theory to problematize these uneasy commitments to either grand theory or postmodernism. In particular, he draws on Susan Buck-Morss's (1989) rendering of Walter Benjamin's massive Arcades Project (Dialectics of Seeing) to re-frame his own spatial materialism. In Benjamin's thought, the movement of history is constructed out of fragments of the past, a view of the city as a kind of shattered urbanism comprising myriad fragments of localized and everyday practices. The result is a critical social theory lodged against what Deutsche (1995:172) astutely called 'the certainties of the singular spatial consciousness which erases the traces of its erasures in a foundational vision of social totality.'

Conversations

At times, Geographical Imaginations is not an easy text and responses to it have varied. For some it is erudite but esoteric, richly theoretical but lacking in concrete or practical engagement (Barnett, 1995; Swyngedouw,

1995; Harvey, 1995). For many readers Geographical Imaginations appeared as a breath of fresh air, an exciting and full-blown encounter with social and cultural theories of all kinds, animated by geographical imaginaries that few at that time had been able to conjure (Deutsche, 1995). Explaining the differences in these responses is not easy and probably has more to do with the tenor of other debates in the mid-1990s than with the success or failings of Geographical Imaginations. At the time, Geography had experienced a decade of theoretical shifts from political economy (Marxism and post-Marxism), to identity politics (feminism, race studies, and gender and sexuality studies) and a poststructural cultural studies committed to a politics of contingency, context, and the concrete.

One of the central goals of the book is to demonstrate the importance and value of close, careful reading. Indeed, at one level, Geographical Imaginations might be interpreted as a response to the readings of Lefebvre that had entered Geography in the 1980s and an attempt to model alternative readings that weren't (or couldn't ever be) closed around a single interpretation. This question of closure has become even more important in his subsequent books, first in his impatience with 'Geography with a Capital G' and second with what in The Colonial Present he refers to as the sense that the world exists in order to provide examples of our theorizations of it. This was a call for an ethic of reading in which authors and their writings are not read in terms of a logic of friends and enemies, of battle and victory, but one in which authors and texts are read for the insights they provide and the work they do, placed in conversation with each other. Gregory is driven by a fear of closure, by that false certainty that leads others to 'know' that answers are clear and pathways are known. Above all, he strives to keep the questions open, to highlight the political commitments of all claims to absolute grounds, fearing that otherwise we close off possible politics-to-come. The violence of the fixed category and the 'disciplined' subject/mind is his target, and this allows him to roam widely among spatial analysts, ethnographers, Marxists, and others without fear or regret that distinct disciplinary-markers were disrupted or new ones not set. For some readers the result has been a deep anxiety about the project of post-structuralism and deconstruction. Swyngedouw (1995: 388) sees this commitment to philosophies of indeterminancy, uncertainty (or better perhaps over-determination) as a laudable goal, but one that leaves political decision and Gregory himself 'up for grabs.' What, Swyngedouw asks, are the practical political commitments that follow from such rich and erudite textual and theoretical analyses?

Gregory begins *Geographical Imaginations* with a suggestion that the book can be read as a reactivation of three dialogues, first with political economy, second with social theory, and third with cultural studies. What does this 'working-through' these dialogues signify and how are we to read the path to cultural studies (Gregory, 1994: 6)? Clive Barnett has suggested that this is fundamentally a book focused on 'theory' and 'space,' and in many ways this is the case. But Gregory reads his own efforts differently as an attempt to articulate more directly a critical, non-metropolitan, spatialized theory of culture *and* a geographically nuanced theory of history.

This is the broader and more interesting conversation with which Gregory asks us to engage. Far from ignoring the political lessons of the Marxist writer Antonio Gramsci, as Katz (1995: 167) suggests, *Geographical Imaginations* can be read as taking up Gramsci's claims about conjunctural analysis in ways that have enormous importance for how we understand geography and the political possibilities of critical theory. In this Gramscian sense, a conjuncture describes 'the complex historically specific terrain of a crisis which affects – but in uneven ways – a specific national-social formation as a whole' (Hall, 1988: 127). But this 'terrain' of conjunctural analysis is not merely contextual. The spatialized and territorialized understanding of context as located and placed must also be understood in terms of the always relational nature of context and place. The reader of *Geographical Imaginations* is thus asked to consider what critical human geography looks like when it takes conjunctural, contextual, and relational analysis seriously and what historical materialism and cultural studies look like when they embody thoroughly spatialized practices.

Conclusion

It is, I think, central to understanding this massive book and what others have at times seen as an unwieldy text whose politics is 'up for grabs,' that we appreciate the ways in which the various finite ontologies and fractured epistemologies Gregory weaves into his analysis enable a certain kind of cultural studies to emerge, one that connects 'the history of the body with the history of space.' The modernist sensibility, the god-trick of grand theory, and the utopian gesture are here displaced in favor of a geography 'that recognizes the *corporeality of vision* and ... requires a scrupulous attention to the junctures and fissures between many different histories: a multileveled dialogue between past and present conducted as a history (or an historical geography) of the present' (Gregory, 1994: 416). And it is here Gregory's reactivation of dialogues works its way into cultural studies. As Larry Grossberg has written:

> Cultural studies is a project not only to construct a political history of the present, but to do so in a particular way, a radically contextualist way, in order to avoid reproducing the very sorts of universalisms (and essentialisms) that all too often characterize

the dominant practices of knowledge production and that have contributed (perhaps unintentionally) to making the very relations of domination, inequality and suffering that cultural studies desires to change. Cultural studies seeks to embrace complexity and contingency, and to avoid the many faces and forms of reductionism. (Grossberg, 2006: 2)

Geographical Imaginations builds a geographer's bridge to this complex and contingent conjunctural history of the present and the multiple spatialities that shape it, and it does so in order to clarify the always political possibilities of such a spatialized history of the present, a cultural politics that becomes even clearer in *The Colonial Present* and *Violent Spaces*. While his critics ask Gregory to more directly address crucial issues of race or

gender, he is already more interested in how we can understand what Stuart Hall (1995: 53–54) – in suggesting that he himself has 'never worked on race and ethnicity as a kind of subcategory' – called 'the whole social formation which is racialized.' That is, the path to cultural studies is about the production of certain kinds of conjunctural truth. These are never fixed, but they intervene in the ways in which social life is produced (Hall, 1997: 157). This is 'a description of a social formation as fractured and conflictual, along multiple axes, planes and scales, constantly in search of temporary balances or structural stabilities through a variety of practices and processes of struggle and negotiation' (Grossberg, 2006: 4-5): a *vital* conversation indeed for geographers and non-geographers alike!

Secondary sources and references

Barnett, C. (1995) 'Why theory?', *Economic Geography* 71: 427–435.

Buck-Morss, S. (1989) *The Dialectics of Seeing: Walter Benjamin and the Arcades Project*. Cambridge: The MIT Press.

Deutsche, R. (1995) 'Surprising Geography', *Annals of the Association of American Geographers*, 85 (1): 168–175.

Gregory, D. (1979) *Ideology, Science and Human Geography*. London: Palgrave Macmillan.

Gregory, D. (1982) *Regional Transformation and Industrial Revolution*. London: Macmillan.

Gregory, D. (1994) *Geographical Imaginations*. Oxford: Blackwell.

Gregory, D. (1995) 'A Geographical Unconscious: Spaces for Dialogue and Difference', *Annals of the Association of American Geographers* 85 (1): 175–186.

Gregory, D. (2004) *The Colonial Present: Afghanistan, Palestine, Iraq*. Oxford: Blackwell.

Gregory, D. and Pred, A. (eds) (2006) *Violent Geographies*. London: Routledge.

Gregory, D. and Urry, J. (1985) *Social Relations and Spatial Structures*. London: Palgrave Macmillan.

Gregory, D., Martin, R., and Smith, G. (eds) (1994) *Human Geography: Society, Space, and Social Science*. London: Macmillan.

Grossberg, L. (2006) 'Does cultural studies have futures? Should it? (Or what's the matter with New York?)' *Cultural Studies* 20 (1): 1–32.

Habermas, J. (1988) *Theory and Practice*. Boston: Beacon Press.

Hall, S. (1988) *The Hard Road to Renewal: Thatcherism and the Crisis of the Left*. London: Verso.

Hall, S. (1995) 'Negotiating Caribbean identities', *New Left Review* 208: 3–14.

Hall, S. (interviewed by David Scott) (1997) 'Politics, contingency, strategy', *Small Axe* 1: 141–159.

Harvey, D. (1989) *The Condition of Postmodernity: An Enquiry into the Origins of Cultural Change*. Oxford: Blackwell.

Harvey, D. (1995) 'Geographical Knowledge in the Eye of Power: Reflections on Derek Gregory's *Geographical Imaginations*', *Annals of the Association of American Geographers* 85 (1): 160–164.

Jones, J.P. (1995) 'Derek Gregory's *Geographical Imaginations*: Dialogic Invitation', *Annals of the Association of American Geographers* 85 (1): 159–160.

Katz, C. (1995) 'Major/Minor: Theory, nature, and politics', *Annals of the Association of American Geographers* 85 (1): 164–168.

Lefebvre, H. (1972/1991) *The Production of Space*. Oxford: Blackwell.

Soja, E. (1989) *Postmodern Geographies*. New York: Verso.

Sparke, M. (2006) *In the Space of Theory*. Minneapolis: Minnesota University Press.

Swyngedouw, E. (1995) 'Geographical Imaginations: Book review', *Transactions of the Institute of British Geographers* NS 20: 387–400.

GEOGRAPHIES OF EXCLUSION (1995): DAVID SIBLEY

Phil Hubbard

> In order to understand the problem of exclusion in modern society, we need a cultural reading of space, what we might term an 'anthropology of space', which emphasizes the rituals of spatial organization. We need to see the sacred which is embodied in spatial boundaries. (Sibley, 1995: 72)

Introduction

The cover of *Geographies of Exclusion* (subtitled *Society and Difference in the West*) is a striking black and white photo entitled 'Woolloomooloo Girl' by the renowned fashion photographer, Henry Talbot. Yet it is clear this is not from a fashion shoot. Taken in an inner city suburb of Sydney in the 1950s, it is of young girl, her hair matted and lank, her clothes simple and plain, clambering over a backyard fence (her backyard?). Whether she is escaping or breaking in is uncertain, but either way we suspect her action is a transgression; she is daring to defy adult authority and crossing the boundary that separates her world from a world that is deemed off-limits. The physical boundary – the fence – is at one and the same time a social boundary, between the ensconced world of childhood and a public space that is defined as adult. Given the photo is of an inner city area, we might also infer that the girl is crossing a class boundary, moving from her working-class territory into the social space occupied by the (middle-class) photographer. Our reactions to this picture may be very different, therefore, depending on our own positionality. Maybe we regard the girl as a threat – a 'feral' street urchin who shows little respect for authority? Or perhaps we admire her for her seeming lack of fear, her sense of adventure, her spirit?

Initially, it is hard to see why this image might have been deemed particularly appropriate for what is an incredibly wide-ranging exploration of the construction of the socio-spatial boundaries that divide us along lines of gender, colour, class, sexuality, age and disability. However, questions of childhood development and socialization are central to Sibley's theorized exploration of what drives us to construct boundaries between those who we feel affinity with and those Others that we regard as different or discrepant. Indeed, psychoanalytical ideas about how a child positions itself in relation to other humans – as well as non-human aspects of the 'object world' (Sibley, 1995: 5) – are at the heart of Sibley's analysis of how difference is created and maintained spatially. Drawing on the work of those psychologists and psychoanalysts who have reworked and extended 'objects relations theory' to consider how we *project* feelings of disgust onto particular people and objects while we *introject* others (which become part of our sense of Self), Sibley's book encourages us all to reflect on the processes by which we become (or became) fully formed individuals, and to think about the exclusions and repressions that are inherent to those processes.

However, the image on the cover is also suggestive of another theme within *Geographies of Exclusion*; namely, the power of the media to perpetuate stereotypes in which purity and pollution are mapped onto specific groups and identities. It is perhaps significant that at the time that Sibley was writing *Geographies of Exclusion*, there was something of a 'moral panic' in the UK about the ways in which teenagers' behaviour often challenged and undermined adult hegemony, at the same time that it endangered younger children. Indeed, another highly charged image in the book is of a pair of teenage boys on a child's climbing frame, glaring at a younger boy. As Sibley relates, these teenagers appear 'out of place' in a space designed for children, yet neither do they appear old enough to access the spaces and sites of adulthood. In the context of a children's playground, they are a transgressive and polluting presence: from an adult or parent's perspective, they might be regarded as 'folk devils' who threaten younger, innocent 'angels'. Sibley relates some of the consequences of this, noting that adolescents were increasingly finding their presence in public spaces regarded as problematic by developers, planners and town centre managers who suggested they threatened the integrity of 'family' consumer spaces (see Vanderbeck and Johnson, 2000). For Sibley, this demonstrates that attempts to order a continuous social category (age) into sharp and neat categories (child/adult) always creates ambiguities and indistinction – and that this has wholly negative implications for those who are represented as existing in these liminal zones.

Consequently, the image of the girl in a Sydney back-alley captures some of the key empirical and theoretical concerns that inform Sibley's book, an inspired intervention in geographical debates about socio-spatial segregation. What is perhaps particularly remarkable about the book, however, is that within just 190 brief pages, it manages to construct an impassioned argument for geographers to adopt more reflexive, engaged and inclusive modes of enquiry, as well as offering a highly original analysis of the way that questions of subjectivity and power construct boundaries at different spatial scales. By drawing parallels between the exclusion of particular knowledges and the exclusion of particular subjects, Sibley was able to produce a book that speaks to a range of debates, marking it out as a text that was of significance well beyond the sub-discipline of social and cultural geography.

Key themes: diversity, difference and defilement

Never the most prolific of geographers, David Sibley's previous publications had nonetheless demonstrated his commitment to a radical and engaged geography. *Outsiders in Urban Society* (Sibley, 1981) was a largely empirical exploration of the lives and spaces of travelling gypsy communities in the UK, based on years of close contact with such groups. A key theme in this work was the way that sedentary society (in the form of local authority councillors and planners) sought to limit this mobile population's residence to 'official' sites, typically located in marginal sites on the edge of cities, away from those wealthy areas of residence where suburban dwellers might complain about the presence of these 'dirty' and 'dangerous' Others. From the perspective of the sedentarist, removing these polluting Others thus 'purified' urban space; conversely, the association of gypsy-travellers with devalued and often derelict sites served to cement the imagined association between their mobile lifestyles and decay and dirt.

Hardly a conventional geographic text, *Outsiders in Urban Society* drew on ongoing Marxist debates about deprivation and class struggle, but sought to extend and amplify these arguments by exploring the different senses of order inherent in both 'mainstream' and 'outsider' society. However, given its focus on gypsy-travellers in the UK, the wider theoretical implications of *Outsiders in Urban Society* remained rather muted, and it was not until a 1988 'review' piece in the journal *Society and Space* that Sibley offered a more explicit account of spatial purification. Herein, Sibley described exclusion and constraint as the outcome of an interplay between processes of individuation (on the one hand) and ideologies promoting property-ownership and capital accumulation (on the other). Underpinned by ideas derived from *structuration* theory, Sibley thus affirmed the importance of transcending structure/agency models in human geography (a widespread concern at the time) through an approach attuned to the reciprocity of individual, society and the environment. Here, he drew on the work of sociologist Anthony Giddens, as well as geographers Allan Pred and Derek Gregory. What was particularly distinctive about Sibley's account, however, was his deployment of the ideas of anthropologist Mary Douglas and educationalist Basil Bernstein. The former had written extensively of purification rituals in tribal societies, highlighting the persistence since ancient times of attempts to impose a symbolic order in which the boundaries between cleanliness and dirt were maintained through both routine and ceremonial practices: The latter had written of the different forms of control evident in the educational sphere, noting the differences between closed and open curricula, where the latter exhibit little concern about boundaries between disciplines or forms of knowledge. From this, Sibley took the idea that we can talk of strongly- and weakly-classified spaces, with

the contemporary urban West principally characterized by strong spatial boundaries, homogeneous spaces and a distinct lack of social mixity. Purification rituals thus entail attempts to maintain this socio-spatial order, removing those who 'pollute' otherwise pure spaces.

By *Geographies of Exclusion*, Sibley (1995: 76) admitted that structuration theory was now regarded as somewhat passé, though continued to insist that it offered a useful way of thinking about the distribution of power. However, he acknowledged a significant critique of structuration theory, namely that it split society into structure (context) and agency (individual intentionality) without exploring the idea that each individual constitutes society differently. As such, Sibley shifted to a theoretical perspective in which structure/agency issues are reframed as questions of the relation between Self/Society. Here, Sibley found the insights of particular packages of psychoanalytical thought useful, not least the distinctive object relations perspective developed by Melanie Klein. Noting that, from the moment of birth, a child experiences 'discomforts of being' (cold, light, noise, etc.), Klein surmises that the child becomes reliant on a mother-figure. However, as the child develops, and recognizes its separation from the maternal (in what Lacan termed the 'mirror stage'), Klein argues that it seeks to redress its sense of loss by bringing some people and things into the Self while rejecting and distancing itself from 'bad objects'. As Sibley notes, the fact that most objects (and people) are not polarized along a continuum of good–bad means that subjects exhibit a great variety of reactions to difference, with some possessing a greater boundary consciousness than others. Hence, while some people are prone to embrace difference, others reject it (and most exhibit a mix of the two tendencies).

Sibley (1995: 7) hence begins from the position that anxieties associated with maternal separation set in motion a series of processes by which individuals seek to construct a boundary between an 'inner (pure) Self and the outer defiled Self'. However, at this point he acknowledges one of the major criticisms levelled at psychoanalytical theory in the social sciences – i.e. its tendency to essentialize and generalize about processes of difference-making – by arguing for accounts more alert to the cultural specificity of boundary-making. Here, he notes that it is 'socialized and acculturated' adults who teach children about the dangers of dirt and pollution, and it is their obsessions and anxieties that transmit to the child. Accordingly, he notes that many of the key anxieties about dirt, soil, faeces and bodily residues that assume so much significance in many instances of boundary-making are not innate, but are distinctly modern and Western (or at least white Northern European) concerns. His subsequent discussion of Julia Kristeva's notion of the abject (defined as that which we seek to distance ourselves from, but remain haunted by) is hence illustrated with examples that are specific to particular geographic place and time (even if some enjoy strong historical resonance).

Sibley's deployment of object relations theory to explain social-spatial issues was by no means the first invocation of psychoanalytic theory by a geographer. However, it was certainly one of the most sustained attempts to explore how 'inner' torments and anxieties manifest themselves in attempts to order the 'outside world'. Moreover, it was also one of the most distinctive, combining Sibley's reading of psychoanalytic theory (Klein, Kristeva, Winnicott, and Erikson in particular) with a range of diverse scholarships drawn from anthropology, history, educational studies, feminism and sociology. In fact, few of the major reference points in *Geographies of Exclusion* are from within geography's established canon. This makes it a potentially challenging read, particularly for those not acquainted with psychoanalytical theories (which were, at least at the time, far from the centre of the geographical curriculum). Yet Sibley's writing style – never verbose and often refreshingly straightforward – provides considerable clarity, and the book itself unfolds through a series of linked chapters which outline how feelings about difference become mapped onto groups regarded as deviant and dirty, and in turn, explain how this produces exclusionary urges whose manifestations are clear at a variety of spatial scales.

Throughout the book, Sibley throws in suggestive examples of the 'imperfect' people who, through history, have been depicted as troublesome Others who need to be located 'elsewhere'. By highlighting instances of xenophobia and discrimination against groups such as prostitutes, gypsies, those with impairments and sexual and racial minorities, he is able to demonstrate that there is a remarkable recalcitrance in the languages and attitudes expressed towards these Other groups in the urban West. Nonetheless, he insists that the boundaries of society are always shifting, and is at pains to illustrate how some groups have drifted into (or conversely, out of) categorizations of purity. The examples that litter the text are consequently many and various, and Sibley draws on a range of popular texts and representations, including films (*Taxi Driver, Simba*), TV adverts (for Persil washing powder, Volkswagen cars) and even the images on a packet of biscuits. These illustrations allow Sibley to vividly support his underlying argument whilst demonstrating the very banal and taken-for-granted nature of many of the representations which figure our reactions to difference.

The first hundred pages or so of *Geographies of Exclusion* hence develop an astute and richly illustrated geographical interpretation of social exclusion. It is at this point that Sibley flips his argument to turn his gaze onto the discipline of geography. Here, his chief contention is that geography (like other disciplines) is based on a hierarchy of knowledge which privileges particular (white, male, middle class) views of the world, but disables and marginalizes other viewpoints. This was not an uncommon argument at the time, given geographers' growing enthusiasm for post-structural theories which are open to multiplicity and difference. Yet Sibley chooses to underline his critique of geography as exclusive by highlighting the way that certain bodies of knowledge have become mere footnotes in geographical histories because their authors peddled a 'localized' knowledge which was viewed as polluting the scientifically elegant and ordered accounts that were regarded as central in establishing geography as a spatial science. Sibley thus alights on the work of W.E.B. Dubois, whose perspectives on the social geographies of the black inner city were dismissed because of his own racial status, and the proto-feminist geographies produced by Florence Kelly and Jane Addams at the Hull House in the University of Chicago at the turn of the nineteenth century. Focusing on these rather obscure figures allows Sibley to suggest that geography is neither attuned to the range of exclusions that fracture the social world nor the exclusions that permeate the discipline – implying that geography itself needs to transcend disciplinary boundaries in order to produce more inclusive and useful accounts of social difference.

A critical assessment

While *Outsiders in Urban Society* had offered a focused case study of social marginalization through spatial process, *Geographies of Exclusion* offered a more wide-ranging and explicitly theoretical overview. Whether it actually offers a cohesive *geographical* theory of exclusion is, however, a moot point – one which

Jonathan Smith (1996: 630) raises in a critical review in which he suggests most readers will ask 'what any of this has to do with geography?' For instance, while Sibley dwells on the importance of stereotyping, and uses multiple media representations to illustrate the pervasive nature of stereotypes of Otherness, it is unclear whether he attributes particular importance to the media as a maker of territorial divisions or whether he is using these examples simply to illustrate the nature of more widespread social anxieties. Further, while Chapters One to Six unfold in a laudable logical progression, moving from a consideration of anxieties around subjectivity through to the ways in which these anxieties are projected onto specific people and places, it remains unclear if Sibley is offering a holistic interpretative framework that might be used by others in their explorations of Otherness, or whether he is simply determined to make a general argument about the connections between Self, society and space. Indeed, in the Introduction, Sibley (1995: ix) modestly claims his intent was 'not to provide a comprehensive account of exclusionary process', merely to illustrate some of the 'more opaque' instances of exclusion that persist in the West.

Further, Sibley's use of Kleinian psychoanalytic theory, while original, may be deemed inappropriate and ill-fitted to considering the wide range of exclusions extant in the urban West (and beyond). Reviewing the book, Cresswell (1997) alleged that Sibley located the construction of difference in a 'natural' transcultural and transhistorical moment, namely, the separation from the maternal. For Cresswell, this led Sibley to assume there is a 'natural' desire to construct boundaries, with the subsequent form of those boundaries being the product of social and cultural forces. Cresswell is clearly troubled by this, and is prompted to ask how the individual anxieties Sibley talks of might be 'scaled up' to the level of the social and cultural. The idea that group behaviour may be no more or less than the culmination of individual behaviours is indeed problematic (and it is notable that Cresswell's own analyses of exclusion and transgression bequeath the media with much more importance in the process of constructing social identities – see, for example, Cresswell, 1996).

Yet even if some commentators regarded Sibley's attempt to integrate psychoanalysis into human geography as not wholly successful, the subsequent citation and uptake of Sibley's ideas suggest that *Geographies of Exclusion* was an important and influential intervention in geographical debates. One factor here was the sheer timeliness of its publication. *Geographies of Exclusion* appeared at a time when talk of social exclusion was to the fore of political and academic agenda, with 15 years of Conservative rule in the UK seen as exacerbating the gap between the haves and the have-nots. Given cities were seen to be fracturing along multiple axes of difference (race, class, age, sexuality, etc.), the class-based heuristics of Marxism appeared increasingly unable to grasp the complexities of the processes that were condemning some to life on 'no-go' estates while others enjoyed unprecedented affluence. As such, the idea that there was a growing group not just marginalized but actively excluded from participation in modern (consumer) society was encouraging many geographers to explore the cultural, social and political basis of inequality, and not just its economic 'underpinnings'.

In this sense, Sibley's thoroughly enculturated interpretation of socio-spatial exclusion appeared at a time when a putative 'cultural turn' was encouraging geographers to explore the role of representation and language in the making of social categories (see Hubbard *et al.*, 2002). Such approaches effectively re-energized social geography's long-standing interest in spatial segregation by considering the cultural politics of marginalization, with notions of 'social sorting' and class differentiation being supplemented by a new-found interest in questions of socio-cultural exclusion, marginalization and its resistance. Related to this was the increasing attention devoted to all number of Others in human geography in the 1990s, with several studies of those living with bodily impairments (e.g. Wilton, 1998), the homeless (Takahashi, 1997), sexual minorities (Hubbard, 1999) and ethnic minority groups (Holloway, 2004) drawing explicitly on the theoretical ideas outlined by Sibley in *Geographies of Exclusion*. Interestingly, Sibley makes repeated reference to representations of humans as animalistic in *Geographies of Exclusion*, presaging a major interest in a human geography of animals (as opposed to zoogeography or biogeography); in later collaborative work he was to focus explicitly on the place of the animal in civilized society (Griffiths *et al.*, 2000), making a significant contribution to the understanding of animals as Others (Wolch, 2002).

Geographies of Exclusion appears highly prescient in other senses, not least its deployment of psychoanalytical theory. At the time, human geography's engagement with psychological theory was largely limited to 'scientific' cognitive-behavioural models of the mind which afforded little importance to the realm of the unconscious, with all that implies about the repression of particular hopes, desires and fears. In his acknowledgements, Sibley credits the geographer Chris Philo for encouraging him to persist with his more 'bizarre' ideas, noting that these were nonetheless becoming quite conventional in geography. Published around the same time as Steve Pile's (1996) *The Body and the City*, and Pile and Thrift's (1995) *Mapping the Subject*, *Geographies of Exclusion* did indeed appear at a moment when human geography was ready to engage with psychoanalytical theories in a serious and sustained manner, with the long-held view that psychoanalytic theory offered little insight into the production of space dissipating in the wake of careful re-interpretation. Sibley's engagement with object relations theory nonetheless distinguishes his approach from some of the deployments of Freud and (especially) Lacan which have been adopted by geographers in fields such as cinematic geography, feminist geographies and the examination of 'emotional geographies' (Aitken and Craine, 2002; Bondi and Davidson, 2004). Significantly, perhaps, Sibley says little about the precarious achievement of gender and sexual identities in and through space – an area in which psychoanalytical geographies have proved particularly insightful (Philo and Parr, 2003). Indeed, when compared with many other geographical takes inspired by psychoanalysis, Sibley presents a decidedly anaemic and disembodied account where there is little attention given to the relation of looks, bodies and surfaces that serve to position us as subjects-in-the-world.

But perhaps the most cutting critique of the book was offered by Smith (1996) when he identified Sibley's book as a somewhat polemic 'countercultural' geography. Suggesting Sibley opposed the 'usual bogies' of depersonalized middle-class suburbanites whose fear, prejudice and obsession quashes the counterculture, Smith (1996: 631) made the case for a geography which attempts the 'hard work' of exploring the antimonies of mixing, tolerance and trust (on the one hand) and excellence, judgement and criticalness (on the other). The accusation here is that Sibley developed an argument designed to suggest that exclusion results from prejudice and irrational intolerance; to the contrary, Smith suggested exclusion can often be justified in quite rational ways and that Sibley had not explored the real-life circumstances in which decisions to exclude some individuals or groups are made. Working through one of Sibley's examples (i.e. the aforementioned exclusion of teenagers from children's playgrounds) Smith argued that the exclusion of older children might be entirely appropriate: even if they do not intend to 'play badly', he suggests they should be excluded because they have the potential to do great harm to younger children because of their relative strength. He continues by asking whether other forms of difference (such as

criminalized drug-use) might be valid. Whether criminals be tolerated and included or shunned and excluded is clearly a problematic debate, but Smith seems to side with the view that, again, exclusion is often justified. Initially, we might read Smith's criticisms as a reactionary response to Sibley's call for mixity and diversity, yet there is a real underlying concern here that nuanced empirical work is needed to explore the social norms and ties that bind specific communities (i.e. their geographies of inclusion) before we attempt to make generalizations about the causes and effects of geographies of exclusion.

However, as a general demonstration of the potential of psychoanalytical theory in human geography, *Geographies of Exclusion* has yet to be bettered. It also continues to have value as a clarion call for geographers to pursue more participatory approaches – even if commentators like Smith were uncertain if it set out a suitably objective framework for exploring socio-spatial relations. In the book, Sibley notes repeatedly that geographers in the 1990s were becoming increasingly attuned to notions of difference, due in part to the de-centring impulses of postmodern theorization (see especially Ed Soja's *Postmodern Geographies* as discussed by Minca, Chapter 16 this volume). Yet he also noted the gaping chasm between their willingness to recognize diversity and the extent to which they were prepared to move beyond their centred, academic viewpoint:

> The social sciences, and human geography in particular, might now be better equipped to challenge xenophobia, racism and other exclusionary tendencies because of a greater intellectual awareness of difference … However … there is still a distance between authors (mainly male) and their subjects. (Sibley, 1995: 184)

Railing against what he saw as an emerging trend for geographers to engage with texts rather than people, Sibley implied geographers 'talked the talk', but seldom 'walked the walk'. Sibley accordingly implored geographers to go out into the world, to engage with people and to produce more inclusive knowledges. This was hardly the most original of arguments (as he acknowledged), and revisited many of the sentiments enunciated in the wake of the 'relevance' debate of the 1970s. And, unfortunately, it could be argued that *Geographies of Exclusion* (unlike *Outsiders in Urban Society*) was not the best illustration of a text written with the close cooperation and participation of the groups Sibley talks of (its sheer scope and range militates against that). Yet read sympathetically, we can view Sibley's book as a continuing provocation to those geographers who keep difference at an arm's length lest their own neatly compartmentalized and institutionalized worldview is disturbed. Facing our own fears, Sibley concludes, may sometimes be an appropriate way of getting to grips with difference.

Conclusion

Though at times idiosyncratic, *Geographies of Exclusion* nonetheless bears repeated reading because of its sheer lucidity. While it does not spell out a particular method for exploring geographical exclusion, it clearly illustrates the value of psychoanalytical packages of thought in explications of social difference. Given this – and the subtext that the discipline of geography needs to be anti-exclusionary – it is clear that the book continues to have relevance within the discipline as a whole, as well as having had a more substantive and obvious imprint on the social and cultural geography that followed it. In many ways, it is disappointing that Sibley did not follow-up *Geographies of Exclusion* with a more focused study that demonstrated how the conceptual building blocks of his theory might be integrated (much of his subsequent work in fact focuses on the problematic boundary between town and country, given all that implies about distinctions between purity and danger – see Sibley, 1997, 2003). Yet for all this, *Geographies of Exclusion* remains a highly suggestive text whose synthetic approach shows the value of a geography that is prepared to transcend established disciplinary boundaries.

Secondary sources and references

Aitken, S.C. and Craine, J. (2002) 'The pornography of despair: lust, desire and the music of Matt Johnson', *ACME, An International E-Journal for Critical Geographers* 1 (1): 91–116.

Bondi, L. and Davidson, J. (eds) (2004) *Emotional Geographies*. Chichester: Ashgate.

Cresswell, T.M. (1996) *In Place/Out of Place*. Minneapolis: University of Minnesota Press.

Cresswell, T.M. (1997) 'Geographies of Exclusion: Society and Difference in the West, by David Sibley' (book review), *Annals of the Association of American Geographers* 87 (3): 566–567.

Griffiths, H., Poulter, I. and Sibley, D. (2000) 'Feral cats in the city', in C. Philo and C. Wilbert (eds) *Animal Spaces, Beastly Places: New Geographies of Human–Animal Relations*. London: Routledge.

Holloway, S. (2004) 'Rural roots, rural routes: discourses of rural self and travelling other in debates about the future of Appleby New Fair 1945–1969', *Journal of Rural Studies* 20: 143–156.

Hubbard, P. (1999) *Sex and the City: Geographies of Prostitution in the Urban West*. Chichester: Ashgate.

Hubbard, P., Kitchin, R., Bartely, B. and Fuller, D. (2002) *Thinking Geographically.* London: Continuum.

Philo, C. and Parr, H. (2003) 'Introducing psychoanalytic geographies', *Social and Cultural Geography* 4 (3): 283–293.

Pile, S. (1996) *The Body and the City.* London: Routledge.

Pile, S. and Thrift, N. (eds) (1995) *Mapping the Subject.* London: Routledge.

Sibley, D. (1981) *Outsiders in an Urban Society.* Oxford: Basil Blackwell.

Sibley, D. (1988) 'Survey 13: Purification of space', *Environment and Planning D – Society and Space* 6 (4): 409–421.

Sibley, D. (1995) *Geographies of Exclusion: Society and Difference in the West.* London: Routledge.

Sibley, D. (1997) 'Endangering the sacred: nomads, youth cultures and the English countryside', in P. Cloke and J. Little (eds) *Contested Countryside Cultures.* London, Routledge, pp. 218–231.

Sibley, D. (2001) 'The binary city', *Urban Studies* 38: 239–250.

Sibley, D. (2003) 'Psychogeographies of rural space and practices of exclusion', in P. Cloke (ed.) *Country Visions.* Harlow: Pearson.

Smith, J. (1996) 'Geographies of Exclusion: Society and Difference in the West, by David Sibley' (book review), *Geographical Review* 86 (4): 629–631.

Takahashi, L. (1997) 'The socio-spatial stigmatisation of homelessness and HIV/AIDS: towards an explanation of the NIMBY syndrome', *Social Science and Medicine* 45: 903–914.

Vanderbeck, R.M. and Johnson, J.H. (2000) 'That's the only place where you can hang out: Urban young people and the space of the mall', *Urban Geography* 21 (1): 5–25.

Wilton, R. (1998) 'The constitution of difference: space and psyche in landscapes of exclusion', *Geoforum* 29: 173–185.

Wolch, J. (2002) 'Anima urbis', *Progress in Human Geography* 26 (6): 721–742.

22 CRITICAL GEOPOLITICS (1996): GEARÓID Ó TUATHAIL

Jo Sharp

Geography is about power. Although often assumed to be innocent, the geography of the world is not a product of nature but a product of histories of struggle between competing authorities over the power to organize, occupy, and administer space. (Ó Tuathail, 1996a: 1)

Introduction

In *Critical Geopolitics* Gearóid Ó Tuathail (1996a) presents a thorough challenge to conventional geopolitics, which he formulates as the theories and practices of statecraft. This monograph has been hugely influential in political geography, pushing forward a critical vision of the subdiscipline influenced by some of the forms of poststructuralism that have driven Geography's 'cultural turn'. The effect of the 'cultural turn' on political geography has generated a move toward non-traditional political geographical knowledges and a concern with the everyday as a valid space of political analysis rather than focusing solely on the formal arena of state politics and international relations. *Critical Geopolitics* has facilitated a challenging of accepted boundaries within the discipline of political geography 'a geopolitical perspective on the field of geopolitics' (Ashley, 1987: 407) as one commentator put it – to examine those relationships that were previously taken for granted.

Critical Geopolitics first examines the tradition of geopolitics through the work of its key proponents at the turn of the twentieth century (e.g. Mackinder, Ratzel and Haushofer). In the central section Ó Tuathail moves to the arguments of critical commentators such as Bowman, Wittfogel, Lacoste, Ashley and Dalby, before moving on to the final section where he exemplifies his own vision of critical geopolitics. The book finishes with a discussion of the changing nature of geopolitics, and especially the influence of technology and media, arguing for the necessity for critical analysis of these new trends: 'The challenge for critical geopolitics today is to document and deconstruct the institutional, technical, and material forms of these new congealments of geo-power, to problematise how global space is incessantly reimagined and rewritten by centers of power and authority in the late twentieth century' (Ó Tuathail, 1996a: 249).

From conventional to critical geopolitics

For Ó Tuathail, conventional geopolitics is understood as an approach to the practice and analysis of statecraft and international relations which considers spatial relations to play a significant role in the constitution of international politics. British geographer and strategist Halford Mackinder popularized the term geopolitics in his famous address to the *Royal Geographical Society*, 'The Geographic Pivot of History', (1904) which promoted the study of geography as an 'aid to statecraft'. Mackinder accordingly proposed that geopolitics offered one such way in which geographers could inform the practices of international relations. Geopolitics studied the ways in which geographical factors shaped the character of international politics. These geographical factors included the layout of continental masses, their physical size, and the distribution of physical and human resources within them. As a result of these factors, certain spaces are seen as either easier or harder to control, distance is seen as influencing politics (because proximity was regarded as potentially leading to susceptibility to political influence), and certain physical features promote security or lead to vulnerability.

Mackinder's best-known geopolitical argument is presented in his 'Heartland Thesis' which insisted upon the importance of the territory in the centre of Asia (the Asian Heartland) in the history of great states. Mackinder believed that whichever state had control over this territory held a more or less impenetrable position and this would provide a powerful position from which to dominate the world. For Mackinder, unless checked by power in the 'outer rim' of territory proximate to the Heartland, the occupying power could quite easily come to control first Europe and then the world.

Ó Tuathail links the rise of geopolitics with the end of European exploration. He illustrates this with the case of Mackinder, who argued that at this point the world had become 'known, occupied and closed' and that it was 'no longer possible to treat various struggles for space in isolation from one another, for all are part of a single worldwide system of closed space' (Ó Tuathail, 1996a: 27). Ó Tuathail thus claims that geopolitics was central to modernist ways of viewing international space. The emergence of this new science of international politics, Ó Tuathail (1996a: 29) argued, produced 'a Cartesian perspectivalist organization of space with a detached viewing subject surveying a worldwide stage whose intelligibility is, for the first time, becoming visible and transparent'. Thus:

> Geopolitics emerged during the last *fin de siecle* as part of an imperial, Eurocentric planetary consciousness. This was a masculinist, ex-cathedra vision of a dangerous world viewed from the commanding heights of governmental and academic institutions. Geopolitics (and the wider discipline) were thus elements in what Martin Jay (1988) has called the 'scopic regimes of modernity,' rational, ordering and controlling. (Heffernan, 2000: 348)

This is an important argument as it offers a critique of suggestions that space was subordinated or invisible in intellectual thought in the twentieth century (an argument most forcefully made by Soja, 1989; see Minca, Chapter 16 this volume). Ó Tuathail argues (1996a: 24) that the enduring influence of geopolitics shows that spatial thinking was indeed influential in twentieth-century political thought, shaping international politics from imperial conquest to the Cold War. Heffernan (2000: 349), however, suggests that Ó Tuathail may have overmade this point, and Ó Tuathail himself has argued the nature of space in geopolitical reasoning is highly reductionist. Conventional geopolitics hence reduces spaces and places to concepts or ideology. The complexity of global space is simplified to units which singularly display evidence of certain political and cultural elements assumed to characterize that place. For instance, geopolitics perhaps reached its height in terms of influence during the 'Cold War' that followed World War II. In that period, American geopoliticians, influenced by Mackinder's Heartland Thesis, worried about the power of the Soviet Union. They accordingly explained the possibilities of Soviet expansion not as a complex political process of adaptation and conflict, but as a result of proximity. The Domino Theory assumed that Soviets, Communists and socialists everywhere 'were, and are, unqualifiedly evil, that they were fiendishly clever, and that any small victory by them would automatically lead to many more' (Glassner, 1993: 239). This led to preemptive military action. US Admiral Arthur Redford, speaking in 1953, for example, argued that an American nuclear strike on Vietnam was essential in order to halt a Viet Minh victory, because that would set off a chain reaction of countries falling to the Communists, 'like a row of falling dominoes' (in Glassner, 1993: 239). The argument went that the United States had to fight and win in Vietnam, for if South Vietnam went communist then automatically, like falling dominoes, Cambodia, Laos, Thailand, Burma, South East Asia, and ultimately, other parts of the world would do likewise. This process would not stop until it reached the last standing domino, the US.

Having outlined conventional geopolitics, the remainder of the book turns to Ó Tuathail's conceptualization of critical geopolitics. Ó Tuathail insists that geopolitics 'does not simply "happen" but is instead practiced by agents at discrete sites of knowledge production, from where it is disseminated and enforced' (Agnew, 2000: 98). Here, Ó Tuathail applies the theories of French philosophers Derrida and Foucault to the 'problematic of geopolitics, the politics of the production of global political space by dominant intellectuals, institutions, and practitioners of statecraft in practices that constitute "global politics"' (Ó Tuathail, 1996a: 185). He exposes the power inherent in any enunciation of spatiality through his transformation of the noun 'geography' to a hyphenated verb *geo-graphy* (literally, from the Greek, earth-writing (*geographien*)). The hyphen is an important literary device as it 'ruptures the givenness of geopolitics and opens up the seal of the bonding of the 'geo' and 'politics' to critical thought' (Ó Tuathail, 1996a: 67). In his rewriting of geopolitics as critical geopolitics, space is power: no description of political events in the world is merely a reflection of some prior or

exterior condition but instead is a 'will to power' – a use of particular geographical representations to construct possible interpretations and limit meaning. There is always a politics to describing the world because there is always a choice of whose descriptions are used.

While traditional geopolitics regards geography as a set of facts and relationships 'out there' in the world awaiting description, critical geopolitics understands geographical orders to be created by key individuals and institutions and then imposed upon the world as frameworks of understanding. Critical geopolitical approaches seek to examine how it is that international politics are imagined spatially or geographically and in so doing to uncover the politics involved in writing the geography of global space. For geopoliticians, there is great power available to those whose maps and explanations of world politics are accepted as accurate because of the influence these have on the way the world is understood. In turn, these understandings have profound effects on subsequent political practice.

Ó Tuathail's deconstructionist approach is not about defining geopolitics but rather understanding how this particular form of knowledge has been used to particular ends. As he puts it:

> How has the term 'geopolitics' been charged with particular meanings and strategic uses within differing networks of power/knowledge? How has it been put to use in differing times and places? The term 'geopolitics' poses a question to use every time it is knowingly evoked and used. (Ó Tuathail, 1996a: 66)

Critical impact and reception

It is difficult to disentangle *Critical Geopolitics* from a number of wider shifts in the nature of political geography in the early 1990s, but there is no doubt that, as an exemplar of critical geopolitical enquiry, it has had a very significant influence on the subdiscipline. Critical geopolitics has included studies of the language of statecraft (for example, Dalby's, 1990b, work on the characterization of the Soviet Union by Reagan's advisors, The Committee on Present Danger) and foreign policy (e.g. Sidaway's, 1998, work on representation of the Gulf) to demonstrate how international politics are imagined spatially and to uncover the politics involved in this process. However, it is not only the geographical imaginations of the powerful that are studied in critical geopolitics: less formal arenas of politics have also been examined. The 'cultural turn' has fully impacted and the definition of 'the political' bringing questions of representation and the politics of identity very much into the heart of study. This has led to the development of a 'popular geopolitics' which has looked at the way in which global and national political identities have been formed through representations of threats in the media, such as Cold War magazines (Sharp, 1996), or films (Sharp, 1998; Power and Crampton, 2005). Despite (or because) of this influence, there have been a variety of critical engagements with critical geopolitics and specifically Ó Tuathail's work.

Probably the most significant and widespread critique focuses on the meaning of 'critical' in Ó Tuathail's formulation. As Agnew (2000: 98) has argued, 'Critical geopolitics is not just an alternative theory to that of geopolitics because it refuses the drive to certainty and objectivity that such a theory requires'. As Ó Tuathail himself notes, critical geopolitics is not an absence of geopolitics but a variation of it. Ó Tuathail too creates geographies when he writes about geopolitics, and his book, like the texts of those he seeks to critique, contains silences and a will to power in his own cartography of international relations. For example, his appears to be a view from nowhere; there is no sense of an embodied critic 'Ó Tuathail' – only

a relentless unveiling and revealing of all geopolitical texts that he encounters (Sharp, 2000a). Ó Tuathail claims his own work is not a totalizing survey of 'geopolitics' but 'a set of contextual explorations of the problematic imperfectly marked by the term "geopolitics"' (Ó Tuathail 1996a: 18). However, his critique does seem to share a number of elements with its object. *Critical Geopolitics* itself is characterized by a Cartesian perspectivalism: 'The object of site here is not so much the landscape or the globe but the array of pre-existing geopolitical *texts* viewed and read by the detached theoretical eye/I' (Smith, 2000: 368). Thus, critics are suggesting that, just as the geopoliticians that come under Ó Tuathail's scrutiny present themselves as all-knowing observers of the world, and predictors of its political future, so too does he stand apart, detached and all-seeing of their works. Ó Tuathail's account of the geopolitical writings of previous and current intellectuals of statecraft is as much a visualization of the 'world-as-exhibition' (Mitchell, 1988) as Mackinder's famous presentation to the Royal Geographic Society was, providing his own heroic narration, not of world domination or prediction, but of the cunning theorist unmasking the powerful statesmen and their advisors.

For Agnew (2000: 98), this denial of the need for 'ontological commitments' is a key limitation of Ó Tuathail's *Critical Geopolitics*. Agnew (2000: 98) wonders what kind of politics might be possible from this perspective and suggests that 'deconstructing the terms and strategies of geopolitics tells us how but not why geopolitical knowledge is contructed where it is and by whom'. Paasi (2000: 284) has similarly suggested that there is a need to move from text and metaphor that had dominated radical political geography to more immediate forms of analysis and intervention. Others, notably feminist critics, have agreed suggesting that the textual focus of Ó Tuathail's critique constrains the nature of politics, instead insisting on a form of geopolitics that is attentive to the embodiment of the people through whose lives it is articulated (Dowler and Sharp, 2001). Thrift (2000) argues for the need for critical geopolitics to think of bodies as sites of performance in their own right rather than as simple surfaces for discursive inscription. Discourses do not simply write themselves directly onto the surface of bodies as if these bodies offered blank surfaces of equal topography. Instead these concepts and ways of being are taken up and used by people who make meaning of them in the different global contexts in which they operate. Various theorists have tried to work Ó Tuathail's ideas through notions of embodiment to produce a more materially grounded critical geopolitics (e.g., MacDonald, 2006), including Ó Tuathail himself (Ó Tuathail, 1996b), and in terms of the bodies of those caught up in the representational practices described in *Critical Geopolitics* (e.g. Hyndman, 2003). Sparke's (1998) ethnographic account of the production of Timothy McVeigh, the 1995 'Oklahoma Bomber', as a political subject, points to the important and complex interrelationships between the production of geopolitical images and their actual impact on people's daily lives.

Feminist critics have noted a more general failure for *Critical Geopolitics* to acknowledge feminist heritage, so that the widening of 'the political' is seen to come from poststructuralist sensibilities rather than feminist arguments about the personal being the political. Indeed, elsewhere it has been suggested that Ó Tuathail's account reproduces geopolitics as an essentially masculinist practice. Ó Tuathail's intellectual history of geopolitical practitioners and critical geopoliticians is certainly a history of Big Men (in order): Mackinder, Ratzel, Mahan, Kjellen, Hausehoffer, Spykman, Wittfogel, Bowman, Lacoste, Ashley, and Dalby. A few women are allowed into the footnotes, but the central narrative is one of the exploits and thoughts of men (see Sharp, 2000a: 363).

The story of geopolitics presented in *Critical Geopolitics* is resolutely male, not just when discussing the masculinist history of geopolitical strategies of elite practitioners, but also the interventions of 'critical geopoliticians'. Sparke (2000) is surprised that Ó Tuathail's theories of vision does not draw upon the pioneering work of Haraway (1988) about the 'God trick' of the all-seeing eye/I of modern science. Moreover, while women were not part of the hallowed intellectual societies which discussed the practised geopolitics, their bodies most certainly were caught up in the resulting political geographies: flows of migrants, representations of other places in geographical imaginations, constructions of 'women and children' to be protected by the state, new social movements and ecological resistances (see Enloe, 1989, 1993). As Enloe (1989: 1) suggests at the beginning of her attempt to make feminist sense of international politics – *Bananas, Beaches and Bases* – 'if we employ only the conventional, ungendered compass to chart international politics, we are likely to end up mapping a landscape peopled only by men, most elite men'.

It is also worth noting that Ó Tuathail maintains a clear distinction between high and low culture, apparently agreeing with the 'intellectuals of statecraft' that he quotes that the High Politics of Statecraft and international relations are beyond the ken of ordinary people, and beyond the effects of the politics of everyday life (Sharp, 2000a). Popular culture is given short shrift, and when it is mentioned, is described as 'largely propagandistic' and offering only 'crude and conspiratorial reasoning' (Ó Tuathail, 1996a: 114 and 121). A variety of work which examines the significance of 'popular geopolitics' has emerged as an alternative (e.g. Sharp, 2000b). The point here is not just that critical geopolitics needs to study a wider range of texts *per se*, but is an intellectual critique of Ó Tuathail's focus on elitist texts.

In many ways, therefore, *Critical Geopolitics* assumes the intellectuals of statecraft are somehow beyond or outside of hegemonic national culture, or that their pronouncements are somehow unaffected by the circulation of ideas and beliefs therein. But, in order to make their arguments believable to their audiences (in many cases, 'ordinary' people, and not just other intellectuals of statecraft), they must refer to concepts and values that have consonance for the population at large, if their support is to be assured. As Ó Tuathail argued in his 1992 paper with John Agnew, 'geopolitics is not a discrete and relatively contained activity confined only to a small group of "wise men" who speak in the language of classical geopolitics' (Ó Tuathail and Agnew, 1992: 194). Simply to describe a foreign policy is to engage in geopolitics and so normalize particular world views. If this is the case, then surely the media – both high and low culture alike – is intimately bound up with geo-graphing the world, as are a range of activities normally described as occurring outwith the sphere of international politics. By forcing apart the 'geo-politics' of the everyday and the 'geopolitics' of statecraft, Ó Tuathail too readily accepts a 'neo-realist' view of state actors as the primary agents in world politics, rather than accepting the fluid and contested nature of hegemonic values and norms (Sharp, 2000a).

A critical assessment

Appearing at a crucial moment in geography's 'cultural turn' (see Hubbard *et al.*, 2002), *Critical Geopolitics* prised open a space for discussions in political geography addressing the role of representation and discourse. In that sense, it remains an important read, mapping out an agenda for subsequent studies. Yet it needs to be read with an eye for its blindspots and occlusions. For example, although Ó Tuathail suggests in his introduction that '[i]dealized maps from the centre clash with the lived geopolitics of the margin' (Ó Tuathail, 1996a: 2), *Critical Geopolitics* does

not contain much examination of resistance to dominant geo-graphings. In the majority of the book, resistance and alternative geo-graphings lie in the textual interventions of critical geopolitics. Marxist critics have been quick to point out alternative forms of action (Smith, 2000), while others have argued for an 'anti-geopolitics' to open up space for forms of resistance beyond the textual. Routledge defines 'anti-geopolitics' as:

> An ethical, political and cultural force within civil society – i.e. those institutions and organizations that are neither part of the processes of material production in the economy, nor part of state-funded or state-controlled organizations (e.g. religious institutions, the media, voluntary organizations, educational institutions and trade unions) – that challenges the notion that the interest of the state's political class are identical to the community's interests. Anti-geopolitics represents an assertion of permanent independence from the state *whoever is in power.* (Routledge, 1998: 245)

Ó Tuathail's division of life into political-effectual and non-political-ineffectual spheres also silences a whole range of people and groups from the operations of international politics. This division of international and domestic politics reinforces the public-domestic spheres division characteristic of patriarchal capitalist society, where women are effectively contained within the mundane space of home and kept from public space, which is the space of politics, change and new possibilities. Ó Tuathail insists that it is important to maintain a certain specificity to the term 'geopolitics' to keep it as a 'careful genealogical approach to the problematic of the writing of global space by intellectuals of statecraft' (Ó Tuathail, 1996a: 143). However, in line with his own arguments that critical geopolitics should be seen as a project which examines to what end particular geopolitical discourses are used, rather than trying to define the term, the effects of this move need to be examined. This argument has been taken up by Matt Sparke (2005) in his examination of the geographing of national identity and international relations. Sparke highlights the necessity for untiring critique:

> the work of describing the graphing of the geo is never done ... it is a reminder of a responsibility to examine other graphings, other geographies, that even avowedly antiessentialist work may have written out of the geo. (Sparke, 2005: xxxi)

Critical Geopolitics is a challenging book which covers difficult concepts and draws on the complex writings of theorists such as Foucault and Derrida. It brought together a new way of thinking about space and power and has had a significant influence on political geography and the discipline of geography itself. That there have been so many critiques of the book, and so many attempts to develop critical geopolitics in new directions, is testament to the ambition of his arguments rather than any limitations of his vision.

Secondary sources and references

Agnew, J. (2000) 'Global political geography beyond geopolitics: review essay', *International Studies Association* 2 (1): 91–99.

Ashley, R. (1987) 'The geopolitics of geopolitical space: towards a critical social theory of international politics', *Alternatives* XIV: 403–434.

Dalby, S. (1990a) 'American security discourse: the persistence of geopolitics', *Political Geography Quarterly* 9 (2): 171–188.

Dalby, S. (1990b) *Creating the Second Cold War: The Discourse of Politics.* New York: Guilford.

Dalby, S. (1991) 'Critical geopolitics: Difference, discourse and dissent', *Environment and Planning D: Society and Space* 9 (3): 261–283.

Dalby, S. (1994) 'Gender and critical geopolitics: reading security discourse in the new world order', *Environment and Planning D Society and Space* 12, 525–542.

Dowler, L. and Sharp, J. (2001) 'A feminist geopolitics?' *Space and Polity* 5 (3): 165–176.

Enloe, C. (1989) *Bananas, Beaches and Bases: Making Feminist Sense of International Relations.* Berkeley: University of California Press.

Enloe, C. (1993) *The Morning After: Sexual Politics at the End of the Cold War.* Berkeley: University of California Press.

Glassner, I. (1993) *Political Geography.* Chichester: Wiley.

Heffernan, M. (2000), 'Balancing visions: comments on Gearóid Ó Tuathail's critical geopolitics', *Political Geography* 19: 347–352.

Haraway, D. (1988) 'Situated knowledges: the science question in feminism and the privilege of partial perspective', *Feminist Studies* 14: 575–599.

Hubbard, P., Kitchin, R., Bartley, B. and Fuller, D. (2002) *Thinking Geographically.* London: Continuum.

Hyndman, J. (2003) 'Beyond either/or: a feminist analysis of September 11th', *ACME* 2 (1).

MacDonald, F. (2006) 'Geopolitics and the "vision thing": regarding Britain and America's first nuclear missile', *Transactions, Institute of British Geographers* 31 (1): 53–71.

Mitchell, T. (1988) *Colonising Egypt.* Berkeley: University of California Press.

Ó Tuathail, G. (1996a) *Critical Geopolitics.* Minneapolis: Minnesota University Press.

Ó Tuathail, G. (1996b) 'An anti-geopolitical eye: Maggie O'Kane in Bosnia, 1992–1993', *Gender, Place and Culture* 3 (2): 171–185.

Ó Tuathail, G. and Agnew, J. (1992) 'Geopolitics and discourse: Practical geopolitical reasoning in American foreign policy', *Political Geography* 11 (2), 190–204.

Ó Tuathail, G. and Dalby, S. (1998) 'Introduction: rethinking geopolitics: towards a critical geopolitics', in G. Ó Tuathail and S. Dalby (eds) *Rethinking Geopolitics*. London: Routledge, pp. 16–38.

Paasi, A. (2000) 'Review of *Rethinking Geopolitics*', *Environment and Planning D: Society and Space* 18 (2): 282–284.

Power, M. and Crampton, A. (2005) 'Reel Geopolitics: Cinemato-graphing Political Space', *Geopolitics* 10: 193–203.

Routledge, P. (1998) 'Anti-geopolitics: introduction', in G. Ó Tuathail, S. Dalby and P. Routledge (eds) *The Geopolitics Reader*. London: Routledge, pp. 245–255.

Said, E. (1978) *Orientalism*. New York: Vintage.

Sharp, J. (1996) 'Hegemony, popular culture and geopolitics: the *Reader's Digest* and the construction of danger', *Political Geography* 15 (6/7): 557–570.

Sharp, J. (1998) 'Reel geographies of the new world order: staging post-Cold War geopolitics in American movies', in S. Dalby and G. Toal (eds) *Rethinking Geopolitics*. London: Routledge, pp.152–169.

Sharp, J. (2000a) 'Remasculinising geo(-)politics? comments on Gearóid Ó Tuathail's *Critical Geopolitics*', in *Political Geography* 19 (3): 361–364.

Sharp, J. (2000b) *Condensing the Cold War: Reader's Digest and American Identity, 1922–994*. Minneapolis: University of Minnesota Press.

Sidaway, J. (1998) 'What's in a gulf?: from the "arc of crisis" to the Gulf War', in S. Dalby and G. Toal (eds) *Rethinking Geopolitics*. London: Routledge, pp. 224–239.

Smith, N. (2000) 'Is a critical geopolitics possible? Foucault, class and the vision thing', *Political Geography* 19: 365–371.

Soja, E. (1989) *Postmodern Geographies: The Reassertion of Space in Critical Social Theory.* New York: Verso.

Sparke, M. (1998) 'Outside inside patriotism: the Oklahoma bombing and the displacement of heartland geopolitics', in Ó G. Tuathail and S. Dalby (eds) *Rethinking Geopolitics*. London: Routledge, pp.198–239.

Sparke, M. (2000) 'Graphing the geo in geo-political: *Critical Geopolitics* and the re-visioning of responsibility', *Political Geography* 19 (3): 373–380.

Sparke, M. (2005) *In the Space of Theory*. Minneapolis: Minnesota University Press.

Thrift, N. (2000) 'It's the little things', in K. Dodds and D. Atkinson (eds) *Geopolitical Traditions: a Century of Geopolitical Thought*. London: Routledge, pp. 380–387.

23 LOGICS OF DISLOCATION (1996): TREVOR J. BARNES

Philip Kelly

[K]nowledge is acquired in many different ways; there is no single epistemology that reveals the 'truth'. To see how knowledge is acquired, we must examine the local context; that is, we must see how knowledge is obtained, used, and verified in a particular place and time. This view is relativist; there are no absolutes because one's knowledge is always 'local' in origin. (Barnes, 1996: 95)

Introduction

If there is one core idea that the reader of *Logics of Dislocation* is intended to take away, it is that there *is* no core idea that should guide the sub-discipline of economic geography (or indeed any project of knowledge production). Barnes' book is an argument against foundations, essences, fundamentals and universals. Instead, he invites students and researchers to think carefully about where they, as individuals, are coming from, how they are constructing their arguments, and the specificities of the contexts they are describing and explaining. *Logics* is, therefore, a call for modesty in our theoretical projects and an attempt to encourage an ongoing conversation between a plurality of perspectives.

Reading *Logics of Dislocation*

Logics is a challenge. Barnes writes exceptionally well, but his prose demands close attention as he engages in tight philosophical arguments only

occasionally leavened by empirical examples. The range of literatures and disciplines touched upon by *Logics* is also quite dizzying: aside from geography, Barnes also engages substantially with economics, sociology, regional science, philosophy, and mathematics. To his credit, however, Barnes is never cleverly obscure just for the sake of it, and always lucid enough to carry the careful reader along with his argument.

While the reader might be tempted to dive right in with the first chapter, it would be a mistake to skip the Preface. Barnes uses those three short pages to explain some important caveats and contexts for the book that follows. In particular, he lays bare his own complicated relationship with the approaches that he will review and critique – a relationship we will examine more closely below.

In the first chapter, Barnes sets out some important concepts that underpin much of the book. In particular, he explains the critical distinction between Enlightenment and post-Enlightenment thought. Enlightenment thinking, emerging as modern science, rose from the mysticism of religious belief in Europe, carried with it four key characteristics:

1 The belief that progress could be achieved through the application of science and rationality – namely the idea that the world could be improved without divine intervention.
2 The belief that human beings are autonomous, self-conscious decision-making units with fixed characteristics and identities.

3   The belief that an order existed in the (natural and social) world, discoverable through rational inquiry free from politics or subjective biases.
4   The belief that universal truths are possible, unchanging in time and space, creating a direct correspondence between ideas and reality.

In short, the Enlightenment project was founded upon a powerful set of '-isms' – universalism, foundationalism, essentialism, and rationalism. Together these provided the ontological basis for modern science, along with all of the power, creativity and destruction that accompanied it. Barnes' focus, of course, is on one small corner of 'modern science' – economic geography. The zenith of scientific approaches to economic geography was in the 1950s and 1960s, when a group of young and mathematically inclined scholars redefined the field, emphasizing the utility of formal models of economic location and behaviour, the expression of such models in the formal language of mathematics, and the use of quantitative data to test and refine them. Variously labelled location theory, spatial science, or the quantitative revolution, this was a set of approaches that influenced not just economic geography, but the discipline as a whole (see Billinge *et al.*, 1984; also Johnston, Chapter 4 this volume).

The 1970s and 1980s saw a strong reaction against these approaches (e.g. Gregory, 1978) and, in economic geography, new debates emerged based upon Marxism, critical realism, locality studies, and flexible production (see Phelps, Chapter 10 this volume). While each of these promised an alternative to the scientific pretensions of the previous generation, Barnes argues that they all failed to live up to the ideals of a post-Enlightenment philosophy – at best, they 'tottered' between an Enlightenment search for certainty and a post-Enlightenment rejection of such a possibility. The chapter then turns to highlight

three approaches that do seem to offer that promise – post-developmentalism, feminist approaches to labour markets, and post-colonialism. In the rest of the book, however, Barnes chooses not to elaborate on these three sets of approaches. Instead, he devotes himself to laying the philosophical groundwork that opens up economic geography to their promise of a post-Enlightenment approach to knowledge production.

In the second part of the book, Barnes sets about dissecting some of the key tenets of Enlightenment thought. In Chapter Two, he demonstrates that conceptions of 'value' in both Marxist and neoclassical economics are rooted in ways of thinking that continually seek essences rather than trying to understand the distinctive way economic value is created in particular places. Barnes argues such processes need to be seen as contextual rather than universal, and are not reducible to a 'final cause'. In Chapter Three, Barnes examines the way human actors are represented in economic theory – usually as rational economic actors ('*homo economicus*') rather than complex multifaceted individuals coloured by social, cultural and political processes. Barnes' larger argument here is that *homo economicus* is flawed not just because he is 'unrealistic', but because he represents a misguided attempt to find universal rather than contextual truths. In both chapters, Barnes carefully takes apart the philosophical assumptions that have guided (often unconsciously) economic geography. He does so, however, not by asserting the superiority of an alternative worldview, but by carefully examining the *internal* logical consistency of existing approaches. While his arguments may be abstract, the type of knowledge production Barnes advocates as an alternative is very much grounded in everyday experience: contextualized, ethnographic and attentive to the richness and variety of economic life.

The third part of the book then moves away from examining *what* various orthodoxies have to say about the world they study, and shifts towards thinking about *who* is saying it and *how* they are convincing others of their arguments. In Chapter Four, Barnes examines the rise of scientific rationality in human geography and looks at the personalities and places involved. He argues that proponents of spatial science seldom applied the scientific method in its purest form, and insists the rise of location theory in economic geography was based upon the sociology of power relations within the academy, rather than the inherent superiority of a new set of ideas. In Chapter Five, Barnes examines the concepts and models used by economic geographers, paying particular attention to the metaphors that help to construct theory. Using the philosopher Richard Rorty as his guide, Barnes argues that metaphors can be useful tools for theoretical debate, but they can constrain our thinking in important ways. Thus, the metaphors widely used in economic geography, such as the gravity model – which assumes that consumers, for example, will be 'drawn' to the most efficient and economical source of a product – should not be seen as revealing fundamental processes, but as aids to our thinking that prove useful from time to time and in particular contexts.

Having examined the models, metaphors and men of economic geography (the protagonists he discusses are indeed all men, although of course not all economic geographers were; see Phelps, Chapter 10 this volume), Barnes turns in Chapter Six to its mathematics – the language of quantification. Here he shows how mathematics, in the form of inferential statistics, is not a medium of ultimate objectivity, neutrality and truth, but is in fact a very partial way of representing the world. Again, Barnes has a philosopher-guide, in the form of Jacques Derrida, whose method of linguistic deconstruction he applies to question the 'certainties' that reside in mathematical descriptions of the world. Here, the ascendance of a particular mode of

analysis in economic geography is shown to rest on decidedly uncertain foundations, and to be largely a reflection of the (local) institutional power wielded by its proponents.

In Part Four of *Logics of Dislocation*, Barnes turns to three individuals to find examples of the type of thinking that he advocates under a post-Enlightenment framework. They are somewhat unusual choices – all white males whose work dates from the mid-twentieth century and none of them contemporary economic geographers – but for various reasons, each holds an important place in Barnes' affections. The first, Piero Sraffa, was an Italian-born Cambridge economist. While Sraffa offered a revised, but still formally modelled, version of several key tenets of neoclassical and Marxist economics in relation to commodity production, he had the vision to leave his models open to the geographical distinctiveness of particular times and places. In resisting the urge to seek universals and foundations, Sraffa was thus a very unusual economist.

Barnes' second exemplar is Harold Innis, a political economist who sought to explain the historical and geographical development of the Canadian economy. Barnes sees Innis' 'staples thesis', in which the development of Canada was closely tied to the geographical characteristics of primary resource extraction, as a case of contextual theory or 'local modelling'. Innis also appeals to Barnes as a political economist who thought seriously about knowledge, reflexivity, language, and communication. In his own empirically grounded work on the forestry sector of British Columbia, Barnes has accordingly pursued Innis' interest in the spatiality of a resource-based economy (see, for example, Barnes and Hayter, 1997).

The third exemplar, Fred Lukermann, was one of Barnes' teachers during his graduate studies at Minnesota, and was something of a subversive at the height of human geography's quantitative revolution in the late 1950s and early 1960s. Lukermann was interested in industrial location but rejected approaches that focused exclusively on either individual locational decisions (in this case by flour and cement millers) or generalized models of rational location patterns. Instead, Lukermann focused on the specific history and geography of these sectors in particular places. More broadly, Lukermann regarded integrative narrative description as a methodology that is contextually sensitive, resistant to the search for root causes, and open to wider cultural, social and political contexts in explaining economic geographies.

Locating *Logics of Dislocation*

We carry around our geographies and histories. (Barnes, 2006: 42)

A central argument of *Logics*, and the reason it finishes with detailed cases studies of three specific thinkers, is that theories are not absolute truths but are always rooted in the personal and societal contexts of their authors and adherents. It is appropriate, therefore, that we look a little closer at where Trevor Barnes himself is coming from. Obviously, ideas and intellectual positions are constructed in complicated ways that are only ever known to the thinker in question. Barnes in particular is not an easy scholar to 'place' – he doesn't slot neatly into any single paradigmatic pigeonhole. Nevertheless, it is possible to discern in *Logics* the traces of a personal intellectual trajectory.

Barnes' early academic training was at University College London in the mid-1970s. Quantitative spatial science and location theory were the mainstream orthodoxies of the day. With a joint degree in geography and economics, Barnes was exposed more than most to the scientific certainties and mathematical language that were then the currencies of both disciplines. For Barnes and the

generation that preceded him, the development of statistical and other technical skills, the impulse to formally model spatial phenomena, and the reduction of human action to a set of rational decision-making processes was at the core of geographical study. Industrial location, urban systems, or patterns of international trade, were seen as products of ordered and logical systems that could be revealed with appropriate methodologies.

The first significant point in Barnes' intellectual trajectory, then, is this early training in the rationalism of economics and an appreciation of geography as a spatial science. It is, as he notes himself in the Preface to *Logics*, his 'temptation' and his 'habit'. Barnes (1996: vi) confesses to being 'deeply smitten by the logic and orderliness of spatial science', but at the same time feeling an urge to undermine these certainties. Indeed, readers of *Logics* will quickly notice Barnes' mode of argumentation in the book is very much a product of formal logical thinking. In that sense the book is very aptly titled – although it seeks to dislocate conventional thinking in various ways, its method for doing so is very much rooted in processes of logical argumentation: Barnes turns upon and dismantles the scientific approach with its own tools. Barnes' fascination with 'scientific' geography is also evident in his writings since the publication of *Logics*, which have provided ethnographic and philosophical insights into the social networks and intellectual underpinnings of geography's 'quantitative revolution' (e.g. Barnes, 2001, 2003). Thus, Barnes is not straightforwardly a critic of Enlightenment thought – his fascination with it represents both an exorcism of a ghost in his own past, and a continued excitement with its technical sophistication and mathematical precision.

To understand Barnes' impulse to deconstruct the scientific economic geography in which he was trained, we need to recognize that while his formal education was focused on quantitative spatial science, his extra-curricular reading reflected the alternative, radical, streams of thought that were emerging in the 1970s through the work of David Harvey and others (see Castree, Chapter 8 this volume). Barnes has noted his debt to Harvey in particular: '[m]y academic life from the time I entered university in 1975 had pivoted around Harvey's works, first, *Social Justice and the City* (Harvey, 1973) that I bought during my first term, and then his various essays appearing in *Antipode*, which as a journal was thought so seditious that it was kept under lock and key at our library. In fact, it was *Social Justice* that persuaded me to attend graduate school' (Barnes, 2004: 408)

At the University of Minnesota, Barnes (2002) recalls stuffing envelopes for the Radical Geography Newsletter, and his doctoral research brought quantitative and radical traditions together in the form of analytical Marxism. This represents a form of radical economics inspired by the categories of Marxist thought, but using sophisticated mathematical tools to substantiate its arguments. Perhaps because of the paradigm-straddling intellectual acrobatics this requires, it has not been a widely adopted approach in economic geography, and Barnes' book with his doctoral supervisor, *The Capitalist Space Economy*, represents one the few examples (Sheppard and Barnes, 1990).

The broader point to note here is that Barnes' training spanned a period of significant change in human geography. What started out as a radical fringe, gradually, over the course of the 1980s in particular, became a central focus of economic geography. Alongside Harvey, the work of Richard Walker, Michael Storper, Neil Smith, Richard Peet, Doreen Massey and others (see Phelps, Chapter 10 this volume; Phillips, Chapter 9 this volume; Coe, Chapter 17 this volume), all made significant impressions. But once again, Barnes' inclination was to cast a subversively sceptical eye upon this ascendant set of approaches. In the Marxist framework that

they apply, Barnes sees a set of logics and certainties just as rigid as those of the spatial scientists they criticized. Thus a core argument of *Logics* is that both Marxist and neoclassical economic approaches suffer from the same problems of essentialism, foundationalism and universalism.

Without entirely burning his bridges with spatial science or Marxism, then, Barnes has carved out a role for himself as the restless analyst of knowledge production in economic geography. It is a role that sits most comfortably with a third set of philosophical influences: post-structuralism. In simple terms, post-structuralism represents an approach to knowledge that rejects any notion of foundations or fundamentals. It argues that there are no absolutes, no 'bedrock notions', no 'God's eye' views of the world. Instead there are partial and contestable knowledges, necessarily situated in the personal experiences of individual knowledge-producers, closely linked to the language of theory construction, and the power relations that surround it (see Doel, 2004).

Logics of Dislocation, then, is Barnes' answer to the puzzle of how one writes a post-structural economic geography that defines the state of the art in the field while at the same time denying the notion that there is a definable field *or* that anyone has the right to try to define it. It is also an exercise in theoretical ecumenicalism − Barnes' purpose is not to debunk 'old' approaches that draw upon neoclassical economics or Marxism, but is instead to strip them of their veneer of certainty and universality in order that they can engage in conversations with each other. Indeed the 'conversation' is an important metaphor for Barnes.

If Barnes' scholarly training bridged the eras of dominance for spatial science and radical political economy in geography, his subsequent home at the University of British Columbia has been a hotbed for the importation of post-structural ideas into human geography. The Department he joined in 1983 included several prominent critics of both spatial science and Marxism, including David Ley and James Duncan; feminist approaches were later represented with the arrival of Geraldine Pratt in 1986 (who had previously been a graduate student in the department). A year later, Derek Gregory, whose trenchant critiques of spatial science and applications of social theory to geography had long influenced Barnes' thinking, also arrived. Perhaps more than any other single department, UBC geographers have collectively brought the post-structural writings of Foucault, Derrida, Haraway, Bourdieu, Said and others into the mainstream of human geography (see Gregory, 1994; Pickles, Chapter 20 this volume). Barnes contributed to, but was also doubtless influenced by, this milieu – and out of it emerged *Logics*.

Logics of Dislocation in context

Having explored Barnes' own flight path towards *Logics*, it is also important to note the intellectual state of economic geography at the time of its publication in 1996. In retrospect, 1995–97 was a fine vintage in economic geography, with several important and lasting contributions. Three contributions marking the vanguard of economic geography appeared almost simultaneously with *Logics*. Susan Hanson and Geraldine Pratt's (1995) influential study of gender and labour markets in Worcester, Massachusetts, had been published the previous year and was cited by Barnes as an example of the 'post-prefixed' economic geography he was advocating. Hanson and Pratt explored the role of space and gender in the constitution of a local labour market without reducing the phenomenon to over-arching frameworks based on class, rational economic decision-making, or gender. They also combined both quantitative and qualitative research methods.

A similarly contextual approach was also being advocated by Jamie Peck in *Workplace* (1996). Peck sought to construct an argument for seeing labour markets as locally constituted through a variety of institutional and regulatory forms. The labour market was thus presented not as a *homogenous* process of class relations under capitalism, nor as the outcome of neoclassical universals based on market mechanisms, but instead as the product of local contingencies. In returning to locality and contingency, Peck's argument was very much in the same spirit as *Logics of Dislocation*. A third significant contemporary of *Logics of Dislocation* was the first edition of J.K. Gibson-Graham's *The End of Capitalism* (1996). Gibson-Graham sought to rethink the ways in which class struggles intersect with other social processes, and to displace the foundational centrality of capitalism in understanding political-economic structures. In their attention to the power of language and representation, and in their non-essentialist focus on multiple processes (class placed alongside gender, sexuality, regionalism, etc.), they too were on the same wavelength as Barnes.

Shortly after *Logics of Dislocation* was published, a collection of research essays by economic geographers titled *Geographies of Economies* appeared and staked out the expanded territories being explored in the field (Lee and Wills, 1997). While Barnes had provided the philosophical arguments for a non-essentialist and pluralist economic geography, this collection showed how economic geographers were operationalizing such an agenda.

The vanguard of economic geography was, therefore, quite closely aligned with Barnes. It is, however, also instructive to consider the state of debate in human geography and the social sciences more broadly. The time at which Barnes was writing, in the early 1990s, was the apogee of controversies over postmodern approaches to knowledge. In the same year as *Logics of Dislocation* was published, for example, a heated exchange occurred in the pages of the *Annals of the Association of American Geographers* concerning a post-modern perspective on the meaning of poverty proposed by Lakshman Yapa (1996). Yapa argued that Indian poverty could be seen as socially constructed through discourses of development and scarcity. Nanda Shrestha (1997: 710) angrily replied that such an argument, focusing on the power of language, amounted to 'aimless intellectual pontification'. Looking back on the debate now, it appears rather overblown, with overstatements and misunderstandings fuelling passions on both sides. A decade later, the idea of social constructivism is well-established and human geography has moved on from the days of 'pomo' and 'anti-pomo' rhetoric. The 1990s, however, were the height of the 'pomo' wars. In some respects, *Logics* reflects this context with its earnest insistence on anti-essentialism and its post-modern rejection of foundational knowledge.

Initial reception and critiques

Logics of Dislocation represented the culmination of almost a decade of work by Barnes. Many of the arguments in the book had appeared in a series of papers published since 1987 and in some cases these papers had themselves stimulated responses and debate (for example, see Barnes 1993, 1994; Bassett, 1994, 1995). The publication of *Logics*, however, brought together Barnes' perspectives on knowledge production in economic geography and laid out the implications for future practice in the field. Reviewers of the book were enthusiastic about the clarity and force of its arguments, but several found themselves reluctant to go all the way with Barnes into the realm of relativist philosophy (for the most thorough consideration of the book, see Bassett, 1996).

Some reviewers raised issues concerning the ways in which the book's arguments were made, without questioning the arguments themselves. For example, it was noted that *Logics* tends to be retrospective rather than prospective. Barnes' focus on spatial science as a target throughout much of the book seemed misplaced given how little influence that paradigm held by the 1990s. His choice of three retired or deceased white male scholars as 'exemplars' also raised eyebrows, given that the first chapter of *Logics* had pointed towards feminism, post-colonialism and post-structuralism as promising ways forward for Economic Geography. These were, however, criticisms that Barnes had anticipated (see pages v–vii), noting that his choice of targets reflected his own past and expertise. Furthermore, Barnes argued that while the Enlightenment thinking that he was dissecting was most explicitly evident in spatial science approaches, the rationalism that underpinned them had lingered in many subsequent frameworks. Some reviewers, however, also accused Barnes of providing slightly caricatured portraits of the theorists he was representing. While Mirowski (1997) suggests Piero Sraffa's work was not quite as sensitively contextual as Barnes portrays, Bassett (1996) conversely argues that Andrew Sayer's critical realism is more subtle than Barnes allows.

A second and more substantive set of criticisms focused directly on the social constructivist approaches to knowledge promoted by Barnes. This critique took four distinct directions. First, while *Logics* advocates a plurality of perspectives, it remains somewhat silent on how we go about choosing between these different understandings of the world. As Bassett points out, Barnes does, on occasion, call for more 'suitable' or 'appropriate' metaphors, and 'a more satisfactory economic geography' (Bassett, 1996: 228), all of which implies that there is some *a priori* basis for making such judgements. But if all knowledge is relative rather than absolute, then how do we determine what is more suitable, appropriate and satisfactory? This also connects with a second criticism, concerning the politics that are implied by Barnes' arguments. Like much of the 'postmodern' philosophy that was being debated at the time, critics suggested that Barnes' approach provided no 'project' – no fundamental beliefs that could serve as a rallying cry for action. Thus while *Logics* advocates feminist, post-colonial and post-developmental perspectives as the most promising ways forward, and all are avowedly political projects, the relationship between theory and political action is not developed in the book. Barnes provides no roadmap for bringing together the 'militant particularisms' of local struggles that would be based on local knowledge. It is important to remember that the Marxism that Barnes criticizes for its essentialism provided a strong dose of politicization for geographers. A third critique concerns Barnes' rejection of grand narratives of knowledge and his insistence that we should use whatever 'works' in a specific context. And yet, Barnes is quite damning of spatial science and Marxism, leaving little room for 'conversations' with them and seeing little value in applying their techniques. It seemed to some, then, that Barnes was himself constructing a grand and linear narrative of paradigm replacement, rather than dialogue. A fourth and related critique asks whether, *in practice*, elements of grand narrative and contextual theory might co-exist, rather than being incompatible as Barnes suggests. Bassett (1996: 456) points out that Andrew Sayer, the critical realist targeted in *Logics*, has argued that 'local knowledges may be appropriate for objects which are indeed local, but grand theories and narratives are needed for objects which are large or widely replicated'. This statement conflates scales of process with levels of abstraction, but the point remains that we still need some of the vocabulary of grand theories when we construct local knowledges.

Logics' legacies: A prelude to new economic geography

Even if *Logics of Dislocation* is just read narrowly as a critical intellectual history of economic geography, it is still a fine achievement. There are few, if any, studies of specific subfields within geography that tackle their subject matter with such a combination of philosophical rigour and ethnographic insight. But *Logics of Dislocation* amounts to much more than a disciplinary history – it is also a manifesto for future directions in the field. It is, however, a very delicate manifesto, which studiously avoids being programmatic or prescriptive. Instead, Barnes' contribution was to lay the philosophical foundations for the 'new economic geography' that was already emerging. We can identify several specific ways in which Barnes' arguments provided a prelude to much of the economic geography that has been conducted since the 1990s.

First, *Logics of Dislocation* explains to economic geographers that they need not seek laws and principles, or read economic landscapes as manifestations of universal processes, but should instead look for processes that are specific in time and space. Barnes argues that explanations of economic geographies *themselves* have a geography – different explanations fit in different places. This recognition of contingency is visible to varying degrees in a range of geographical research over the last decade, from the tracing of commodities chains and networks (Leslie and Reimer, 1999), to examining the role of geographically specific institutions (Peck, 1996), to thinking about embeddedness (Hess, 2004).

Secondly, Barnes' arguments for non-essentialism endorsed a range of approaches that rethought what is meant by 'the economic' and attempt to incorporate a wider range of social processes into studies of economic life. Economic geographers have been particularly active in seeking to incorporate cultural practices and the multiple subjectivities (gendered, racialized, regionalized, etc.) occupied by economic actors. Thinking about economic actors in non-essentialist ways has thus opened up a wide range of possibilities, including the fields of labour geography, ethnic economies and feminist labour market studies.

Third, the methodological corollary of contextualism and non-essentialism is often ethnographic analysis, based on close observation of the ways in which economic processes are actually practised. These processes might include the tacit knowledge and networks that lead to regions of intense technological innovation, or the workplace processes and performances that lead to certain patterns of gendered segmentation in the labour force (see Saxenian, 1996; McDowell, 1997). To capture these processes economic geographers have increasingly used qualitative methods in their research.

Fourth, Barnes' advocacy for reflexivity in the research process and in the construction of explanations has also been widely taken on board. This implies not just thinking about why we choose to ask certain questions as individual researchers, but also to think about the politics of our research and where it stands in relation to social power relations. It is now an expected component of methodological rigour to reflect upon the 'positionality' of the research in the research process (e.g., see Tickell *et al.*, 2007).

Conclusion

It would be an overstatement to suggest that *Logics* brought about all these new approaches in economic geography – most were already apparent when the book was published. What can reasonably be claimed, however, is that the book laid a philosophical foundation for a new approach to knowledge production in economic geography, and much of the research in the field that has appeared since has been consistent with this vision. Indeed, through a series of edited collections published since *Logics*, Barnes has continued to chart (still ever so delicately) the new directions in the field (Sheppard and Barnes, 2000; Barnes and Gertler, 1999). For students seeking to understanding some of the philosophy behind the contemporary practice of economic geography, and where the field has come from, *Logics of Dislocation* remains their best guide.

Secondary sources and references

Barnes, T. (1993) 'Whatever happened to the Philosophy of Science?' *Environment and Planning A* 25 (3): 301–304.

Barnes, T. (1994) '5 ways to leave your critic – a sociological experiment in replying', *Environment and Planning A* 26 (11): 1653–1658.

Barnes, T. (1996) *Logics of Dislocation: Models, Metaphors and Meanings of Economic Space.* New York: Guilford Press.

Barnes, T.J. (2001) 'Lives lived, and lives told: biographies of geography's quantitative revolution', *Environment and Planning D – Society and Space* 19: 409–429.

Barnes, T.J. (2002) 'Critical notes on economic geography from an aging radical. Or radical notes on economic geography from a critical age', *ACME: An International E-Journal for Critical Geographies* 1: 8–14.

Barnes, T.J. (2003) 'The place of locational analysis: a selective and interpretive history', *Progress in Human Geography* 27: 69–95.

Barnes, T.J. (2004) 'The background of our lives: David Harvey's *The Limits to Capital*', *Antipode* 36: 407–413.

Barnes, T.J. (2006) 'Between deduction and dialectics: David Harvey on knowledge', in N. Castree and D. Gregory (eds) *David Harvey: A Critical Reader*. Oxford: Blackwell, pp. 26–46.

Barnes, T.J. and Hayter, R. (eds) (1997) *Troubles in the Rainforest: British Columbia's Forest Economy in Transition*. Victoria: Western Geographical Press.

Barnes, T. and Gertler, M. (1999) *The New Industrial Geography: Regions, Regulations and Institutions*. London: Routledge.

Bassett, K. (1994) 'Whatever happened to the Philosophy of Science – some comments on Barnes', *Environment and Planning* A 26 (3): 337–342.

Bassett, K. (1995) 'On reflexivity – further comments on Barnes and the sociology of science', *Environment and Planning* A 27 (10): 1527–1533.

Bassett, K. (1996) 'Reconstruction or Dislocation: Barnes or Sayer on the Political Economy of Space?' *Geoforum* 27 (4): 453–460.

Billinge, M., Gregory, D. and Martin, R. (1984) *Recollections of a Revolution: Geography as Spatial Science*. New York: St Martin's Press.

Doel, M.A. (2004) 'Poststructuralist geographies: the essential selection', in P. Cloke, P. Crang and M. Goodwin (eds) *Envisioning Human Geographies*. London: Arnold, pp.146–171.

Gibson-Graham, J.K. (1996) *The End of Capitalism (as we knew it): A Feminist Critique of Political Economy*. Cambridge, MA: Blackwell.

Gregory, D. (1978) *Ideology, Science and Human Geography*. New York: St Martin's Press.

Gregory, D. (1994) *Geographical Imaginations*. Oxford: Blackwell.

Hanson, S. and Pratt, G. (1995) *Gender, Work and Space*. London and New York: Routledge.

Hess, M. (2004) '"Spatial" relationships? Towards a reconceptualization of embeddedness', *Progress in Human Geography* 28 (2): 165–186.

Lee, R. and Wills, J. (eds) (1997) *Geographies of Economies*. New York: Arnold.

Leslie, D. and Reimer, S. (1999) 'Spatializing commodity chains', *Progress in Human Geography* 23 (3): 401–420.

McDowell, L. (1997) *Capital Culture: Gender at Work in the City*. Oxford: Blackwell.

Mirowski, P. (1997) 'Logics of dislocation: Models, metaphors, and meanings of economic space', *Canadian Geographer* 41 (3): 335–336.

Peck, J. (1996) *Work-Place: The Social Regulation of Labor Markets*. New York: Guilford Press.

Saxenian, A. (1996) *Regional Advantage: Culture and Competition in Silicon Valley and Route 128*. Cambridge, MA: Harvard University Press.

Sheppard, E. and Barnes, T.J. (1990) *The Capitalist Space Economy: Geographical Analysis After Ricardo, Marx and Sraffa*. London: Unwin Hyman.

Sheppard, E. and Barnes, T. (2000) *A Companion to Economic Geography*. Oxford: Blackwell.

Shrestha, N. (1997) 'On "What causes poverty? A postmodern view": A postmodern view or denial of historical integrity? The poverty of Yapa's view of poverty', *Annals of the Association of American Geographers* 87 (4): 709–716.

Tickell, A., Sheppard, E., Peck, J. and Barnes, T. (eds) (2007) *Politics and Practice in Economic Geography*. London: Sage.

Yapa, L. (1996) 'What causes poverty? A postmodern view', *Annals of the Association of American Geographers* 86 (4): 707–728.

24 HYBRID GEOGRAPHIES (2002): SARAH WHATMORE

Sarah Dyer

... the hybrid invites new ways of travelling. (Whatmore, 2002: 6)

Introduction

Hybrid Geographies is a challenging book. Its theoretical and empirical scope is wide ranging and impressive. It draws on and develops a number of different areas of literature to locate them firmly within the discipline of geography. In this book, Whatmore challenges us to fundamentally rethink the ways in which we understand nature and the natural world. In exploring our conceptualizations of nature we are required to think through the politics of nature, to traverse history and the globe, and to question our assumptions about the relationship between humans and the natural world. This is also a book that challenges many other assumptions. It occupies many contradictory, or seemingly contradictory, positions at once. It is poetic and yet factual, imaginative but not imaginary. The book takes as its subject the 'fleshiness' of the material world and yet it is an intricately theoretical text. It is, as Whatmore (2002: 6) herself says, philosophical but not philosophy. Moreover, it is neither a conventional monograph with a beginning, middle and end, nor a book of essays. It is, again in her own words, 'an effort to germinate connections and openings that complicate' (Whatmore, 2002: 6). A book which has the express aim

of complication can seem a little unnerving – especially for the undergraduate reader. Consequently, it is a book that can be as confounding as it is pleasurable.

In this chapter I aim to provide readers an entry into the text and an understanding of why this text has become a key text in human geography. I begin by giving an overview of the book, describing the contemporary relevance of the topics addressed in *Hybrid Geographies* and geographers' interest in these. I then turn to the main arguments Whatmore makes and the evidence she draws on to make them. I highlight concerns raised by some about the truth status of the claims made in the book. In the next section I describe the acclaim with which the book was received at the time of publishing. I summarize three key questions discussants of the book raised in a conference session about *Hybrid Geographies*. These provide a thought-provoking platform from which readers can perform their own critical reading. I then end with a short consideration of the place *Hybrid Geographies* has within a more unified human-and-physical geography.

Overview

The book begins by invoking a set of debates we are used to seeing on the front pages of our newspapers and on the evening news. In any given week, the media covers many stories which concern our changing relationship with the natural world. These stories include concerns over farming and the food we eat, such as the 'crisis' in the UK in 2001 over the spread of foot and mouth disease or the ongoing debates about genetically modified crops (GMOs). Watching the evening news we are confronted with stories about the risks wild animals pose to humans. The spread of bird flu from 2003 onwards, the contraction of the disease by people, and the subsequent debate about the likelihood of human-to-human infection, is typical. We also see the converse concerns, about the threats humans pose to the natural world, be they marked by extinction rates of endangered animals or rising sea levels. Similarly, stories abound of the new and challenging ways in which science and technology are enabling us to alter our own (human) nature, from extending fertility to enhancing of our capabilities through drugs. As the popular press recognize, these stories and the issues raised are of great concern to people. We are challenged to think about what we can and can't control, how we should behave, and whose responsibility it all is.

Whatmore's premise is that geography brings important tools to the understanding of these issues. In order to introduce her readers to hybrid geographies, Whatmore (2002: 4–6) describes the main bodies of work with which she engages. These are variously labelled ones of 'life politics' or 'bio-sociality' as they involve a reconfiguring of socio-political life, the material world, and the relationships between the two. Geographers have been interested in these issues and these literatures long before *Hybrid Geographies*. However, Whatmore's project is a creative synthesis of existing literatures and to locate them firmly within geography. Elsewhere, Whatmore has argued that social science often ends up following the agendas and lead of the life sciences:

> ... not enough energy is being invested in the much harder and less eye-catching work of contextualising the biotechnologies that are making the headlines in the variegated histories and geographies through which life and knowledge has been co-fabricated. (Whatmore, 2004: 1362)

Geography has the ability to transcend the 'science wars' that pit social against natural sciences, she argues, and to enable social scientists to contribute fully to debates about 'life politics'.

In order to demonstrate the value of geography, Whatmore works through a number of examples in *Hybrid Geographies*. The book is divided into three sections: bewildering spaces; governing spaces; and living spaces. Each section has its own short introduction and a couple of chapters giving what can be thought of as case studies. Each case can be read as self-contained, as illustrating the theme of the section it is in, or as furthering the arguments of the book as a whole. She sets out her agenda and theoretical underpinning in the first and last chapter of the book. I will not summarize the case studies here, as I could not do justice to Whatmore's rich and thorough descriptions. I will turn instead to a wider consideration of the argument the book makes.

Arguments and evidence

Hybrid Geographies is, at its core, a book about relations; it is about the relationship between the social and the natural; it is about the relationships of long-dead leopards that fought in the Roman amphitheatres and of elephants in British zoos; it is about the relationships that construct each of these creatures and, through juxtaposing the stories of these animals in this book, Whatmore poses questions about the relationships between these creatures. It may seem odd to talk about animals having relationships, but that is

Whatmore's starting point. She sets out in this book to challenge conventional ways of thinking about nature and to challenge us to think about the relationships between society and nature. It is only by challenging such conventional dualist thinking we have the chance of honestly representing and acting in the world.

The nature-society dichotomy is a long-standing and powerful one. For a long time (in the Western world) these two realms have been thought to be ontologically separate, mutually exclusive. It is associated with other dichotomies under the labels Modernity or the Enlightenment. Something is social or natural, active or passive, agent or acted upon. In this schema nature is separate from humanity and humans have the monopoly on knowledge, agency and morality. Humans are subjects in the world. They are able to know and to manipulate nature, which is passive and object. Being separate does not mean located in different places, although that too is seen to be true (rural versus urban), but separate ontologically, in the most fundamental way. Whatmore is by no means alone in her attempt to describe an alternative mode of thought. However, the fundamental premise of *Hybrid Geographies* is that seeing the world as being divided up into either the natural or the social (and all the other associated dichotomies) is neither useful nor honest. The fundamental aim of the book is to enlarge the notion of social and to de-couple the subject–object dichotomy (Whatmore, 2002: 4): to write of what Whatmore has called elsewhere, a 'more-than-human geography' (Whatmore, 2003: 139).

It is important to understand that this is not just a book about the relations of nature and society but a book that asks us to understand the world as constituted by relations. Rather than being 'pure' or 'discrete' and belonging to the natural rather than the social entities, things such as elephants or soy beans are constituted through relations. They

are hybrids. Our world is a world constituted through hybridity. And the same is true for people. Although we might assume humans obviously belonging to the realm of the social, as Whatmore shows when she discusses the colonization of Australia, people have always managed to construct a division between social and natural, which excludes some people. Having posited the world as relational, the author provides a tool for making sense of it: hybrid geographies.

In order to understand the relationships that constitute things, the author insists, we must follow the journeys these things take. These might be literal journeys, such as the transportation of leopards to Roman amphitheatres, or the movement of captive-bred crocodiles into the wild. They might also involve translation into information, such as the drawing up of a statute or the monitoring of captive animals' breeding stock, and the movement of this data. Consequently, in this book we travel back and forward in history and traverse the globe. Whatmore conjures up 'wormholes', holes between the past and the present, the distant and the local, the strange and the familiar. These are not imaginary journeys (although they do seek to be imaginative ones), neither is this book simply a report of empirical work. The author seeks to present us with 'closely textured journeys' (Whatmore, 2002: 4). They encompass both the very personal (for example, invoking the moment she became a member of staff at UCL or getting lost on a car journey in Australia) and the more common academic representations of the world. The journeys the people and animals in the book take (literally and figuratively) constitute them.

Whatmore is magpie-like in her collecting of evidence, which she deploys to construct these geographies using both words and images. Using plans of the journeys these animals take (for example, Figure 2.4 on page 25), description of the bureaucratic discourses, and 'thick' descriptions of the spaces these animals inhabit, she paints compelling pictures of their geographies of hybridity. She illustrates how what we think of as 'nature' does not pre-exist the spaces and relationships which it occupies.

Whatmore's use of evidence in her argument has attracted some attention. As Duncan (2004: 161) has noted, the empirical material in the book is 'marshalled as illustrations' in a methodological, theoretical, and ethical project. While this is not in itself necessarily problematic, there is no conventional methods section which lays out the methods of selecting, collecting, and analyzing empirical data. This makes it hard for the more conventionally minded social science reader to judge the merits of evidence and arguments with which they are presented. Braun (2005: 835) sees *Hybrid Geographies* as 'insistently empirical … (that it) seeks to explain nothing more than what is immediately at hand'. While he is sympathetic with Whatmore's approach, he suggests her own knowledge claims, which are presented as detached and objective, do not reflect her professed claim that knowledge is relational and precarious. Demeritt (2005: 821) agrees: 'The epistemic modesty about partial and situated knowledge is somewhat belied by some quite strong claims about how that world actually is.' Conventionally, a methodology offers readers an account and defence of the mechanics of the research project(s). Such an account recognizes the precarious and inter-subjective nature of social science truth claims.

The methodological commitments demonstrated in *Hybrid Geographies* are fundamental to many geographers' vision of the ways in which the discipline ought to develop. Whatmore has since argued that we need to be risky and imaginative in the methods we use:

> (O)ne of the greatest challenges of 'more-than-human' styles of working as I see it is the onus they place on experimentation … (There) is the urgent need to supplement the familiar repertoire of humanist methods that rely on generating talk and text with experimental practices that amplify other sensory, bodily and affective registers and extend the company and modality of what constitutes a research subject. (Whatmore, 2004: 1362)

Certainly, *Hybrid Geographies* meets this challenge although, as noted above, it may be in a way that some geographers find a little opaque. In a paper on Australia, Instone (2004) considers the book's methodological lessons. She argues (Instone, 2004: 134) that Whatmore has challenged geographers to focus 'on the practices through which nature is manifested in social action'. Such a relational analysis requires, she argues, a 'multilayed, multivalent, embodied and situated approach' (Instone, 2004: 131), with 'connectivity at its heart' (Instone, 2004: 136). Such an approach will include textual, visual, oral and/or material evidence. These methods force geographers into 'a process of figuring what matters' (Instone, 2004: 138).

The multilayering and 'risky-ness' of the methods espoused in *Hybrid Geographies* are mirrored by the prose. In the preface to *Hybrid Geographies* Whatmore (2002: x) says she has tried to 'hold onto some sense of this energetic fabrication in the writing'. This emphasis on opening up arguments and connections rather than closing them down or settling them, can lead to rather dense text. However, it is an important strand in her argument. Whatmore (2005: 844) says she sets out to perform her philosophical commitments rather than merely state them. She wants to open up new ways of thinking about nature and to make these imaginatively as well as rationally plausible.

This leads to a style that some find delightful but others find frustrating. It allows the reader, and sometimes requires them, to be playful. For example, there is no exclusively linear argument in the book. This book presents the reader with a robust structure, albeit a malleable one. The three sections structure although there is not necessarily a lateral progression of the arguments through the book. At its core, though, as Demeritt (2005: 882) notes, this is still at its heart an academic text which follows the appropriate scholarly conventions such as citation. For, in a sense, the beginning chapters counter conventional views of nature as apart from the social, while the later chapters provide an alternative construction. Although the final chapter is not a conclusion to the book (it does not rehearse the arguments or sum up the findings), it returns to the aim set out at the beginning of the book to enlarge our notion of the social.

Impacts and significance

Hybrid Geographies received much interest and acclaim at the time of publishing. It was described variously as 'splendid book, brimming over with interesting ideas' and with 'compelling' arguments (Duncan, 2004: 162) and marking 'an important and impressive milestone' (O'Brien and Wilkes, 2004: 149). It was welcomed as '(r)einvigorating geography and making it relevant to interdisciplinary work' (O'Brien and Wilkes, 2004). Geographers of a certain bent had been working on the questions and literatures of science and technology studies for a while. They embraced *Hybrid Geographies* as a book that works through the implications of geographic thought for these concerns and promises the contribution geography can make to this academic area. In the years since its publication the book has been much cited, becoming what can be termed, in ANT parlance, an obligatory passage point, it is a necessary point of reference for many geographers of nature and informs their thinking on the relational co-constitution of nature and society.

In 2003, Noel Castree organized an author-meets-critic discussion at the annual conference of the UK's Royal Geographic Society (RGS). Those involved in this exchange published their thoughts in *Antipode* (Demeritt,

2005; Philo, 2005; Braun, 2005; Whatmore, 2005). These papers provide thoughtful and sympathetic critique of the book. Braun (2005: 835) describes it as a book that 'believes in the world'. Demeritt (2005: 818) calls it fascinating. Philo (2005: 824) characterizes it as an 'excellent book, which seriously moves on debates'. All of the contributions recognize that the book is challenging and provide entries into, and questions to enrich, our reading of the *Hybrid Geographies*. Having already described the concerns expressed in part in this exchange about the status of the truth claims Whatmore makes, I am going to concentrate here on three further questions. The first is Demeritt's concern about how far we ought to extend Whatmore's analysis of hybridity. The second concerns Braun's calls for a clarification of the relationship between knowledge and ethics/politics. Finally, I describe Philo's anxiety about the place of animals in the text.

The first question posed in the *Antipode* exchange concerns the scope of Whatmore's analysis. I have said that at its core, *Hybrid Geographies* is a book about relations. One turn of phrase Whatmore (2002) uses in this book is that of 'becoming…', for example, she talks of becoming a leopard and becoming a soybean. In doing so, Whatmore seeks to highlight the networks and journeys that co-construct leopards and soybeans. While we might be convinced that this is a useful way of thinking about the capacities of animals, we ought to ask is it a useful way of thinking about their fleshiness. Demeritt (2005: 829) asks whether we want to accept an elephant's skin and trunk are partial and provisional achievements. He accepts such a critical realist challenge makes assumption about what partial and provisional mean. However, this is an important question to ask in our reading of *Hybrid Geographies*. While Whatmore makes a compelling case for the socio-natural hybridity of some aspects of nature, how far does the scope of her argument extend?

The second of the *Antipode* questions concerns the place of animals in this text. Although this is a book about animals, there is a sense in which they are absent, or at least veiled, in the text. As Philo puts it:

> ... might it not be that the animals – in detail, up close, face to face, as it were – still remain somewhat shadowy presences? They are animating the stories being told, but ... they stay in the margins. (Philo, 2005: 829)

Although she discusses corporeal 'fleshiness', her examples are drawn from bureaucratic and all too human discourses. As another commentator has said:

> Although at times ... we may occasionally be tempted to ask where the elephants are, in some ways this is the point here – that there are many aspects that define an animal, what it can do, or will be made to do, which are constituted in wide circulations of materials moving in and beyond its fleshy body, though admittedly these can be somewhat abstracted forms here. (Wilbert, 2004: 91)

Duncan too notes how the examples Whatmore works through effectively illustrate how social networks interpenetrate the natural (Duncan, 2004: 162). He wonders what hybrid geographies of animals that are less enrolled in human projects would look like and posit that they would be less human-centred granting greater agency to animals.

The final consideration raised by the *Antipode* exchange that I will mention here is the construction of ethics. The project of ethics is central to *Hybrid Geographies*. In an earlier paper, Whatmore lays out the basis of the relational ethics which makes up the 'conclusion' of *Hybrid Geographies* (Whatmore, 1997). Despite the centrality of ethics to the *Hybrid Geographies* project, there is a sense in which it remains difficult to get a handle on the resultant ethics. Braun (2005: 838) has

argued that 'the question of knowledge and its relation to politics is not always clear in *Hybrid Geographies*', noting in particular the absence of a mention of capitalism in her analysis. Wilbert (2004: 92) has described relational ethics as rather vague. There is a sense in which it is unclear what difference hybrid geographies make ethically. What can we now understand, now do, that we couldn't before? For example, conventional humanist ethical theories take human agency and subjecthood to denote a capacity for responsibility. Whether newly-enlarged hybrid geographies of agency and subjectivity bring with them such moral responsibility is something on which Whatmore remains silent.

The exchange in *Antipode* raises these questions as part of a thorough engagement with Whatmore's book and against a background of praise for its ambitions and execution. They provide useful questions to enrich our own readings of *Hybrid Geographies*. The extent to which these concerns are judged important or worked through rests not with Whatmore herself but with the extent to which the text proves useful to geographers, other academics, activists and beyond. As Whatmore (2005: 842) herself notes, 'books have lives of their own – proliferating connections as they travel that exceed any authorial intentions'.

Conclusion

Hybrid Geographies has been received as an important and challenging book. It has already become a classic among a community of geographers dealing with the relationship between nature and society, already widely cited. To some working in this area, though, its analysis is too apolitical to be adopted. For others, its dense prose and flouting of important social science convention (such as an account of methods) make it inaccessible.

However, despite such reservations *Hybrid Geographies* has become known as an important and agenda-setting book.

This is a book that firmly positions geography as relevant to important issues facing the world and the academic treatment of these issues. In a sense, this will be its acid test. The challenge is twofold. There is the need for geographers to communicate with those beyond geography. There is also the challenge of creating methods and language for geographers to communicate with those in the discipline working in areas other than their own. Elsewhere, Whatmore (2003) has discussed the ability geography ought to have to make sense of biological processes, such as diseases and micro-organisms, as much as political ones, such as international trade agreements. She gives the case of banana production and its threat by a virulent leaf disease called Black Sigatoka:

> This obscure airborne fungus put the banana in the headlines in European news papers, heralding generic engineering as the banana's only salvation. Geography should be a discipline that fosters skills to deal as effectively with this pathological process as with the organisational relations of banana production, but seems less inclined to do so. (Whatmore, 2003: 139)

If Whatmore's book *Hybrid Geographies* can prompt such research agendas, and if it plays a role in the conversation between human and physical geography, it will be a real success. The book will be creating its own hybrid networks. It is certainly up to the challenge.

Secondary sources and further reading

Braun, B. (2005) 'Writing geographies of hope', *Antipode* 37 (4): 834–841.

Duncan, J. (2004) 'Hybrid Geographies: Natures, Cultures, Spaces' (Book review), *Social and Cultural Geography* 5 (1): 160–162.

Demeritt, D. (2005) 'Hybrid geographies, relational ontologies and situated knowledges', *Antipode* 37 (4) 819–823.

Instone, L. (2004) 'Situating Nature: on doing cultural geographies of Australian nature', *Australian Geographer* 35 (2): 131–140.

O'Brien, W. and Wilkes, H. (2004) 'Hybrid Geographies: Natures, Cultures, Spaces' (Book review), *The Professional Geographer* 56 (1): 149–151.

Philo, C. (2005) 'Spacing lives and lively spaces: partial remarks on Sarah Whatmore's Hybrid Geographies', *Antipode* 37 (4): 824–833.

Whatmore, S. (1997) 'Dissecting the autonomous self', *Environment and Planning D: Society and Space* 15 (1): 37–53.

Whatmore, S. (2002) *Hybrid Geographies:* Natures, Cultures, Spaces. London: Sage.

Whatmore, S. (2003) 'From banana wars to black sigatoka. Another case for a more-than-human geography', *Geoforum* 34 (2): 139.

Whatmore, S. (2004) 'Humanism's excess: some thoughts of the "post-human/ist" agenda', *Environment and Planning* A 36 (8): 1360–1363.

Whatmore, S. (2005) 'Hybrid Geographies: Author's responses and reflections', *Antipode* 37 (4): 842–845.

Wilbert, C. (2004) 'Hybrid Geographies: Natures, Cultures, Spaces' (Book review), *Area* 36 (1): 91–92.

25 CITIES (2002): ASH AMIN AND NIGEL THRIFT

Alan Latham

Cities have to be seen less a series of locations on which categorical attributes are piled, and more as forces and intensities that move around and from which, because of their constant ingestions, mergers and symbioses, the new constantly proceeds... Life in the city contains magical powers; it is full to brimming with an abundance of life... (Amin and Thrift, 2002: 91)

Introduction

Urban geography and urban studies is a discipline populated by Big Things. Cities for a start. They are by definition big. Motorways and mass transportation systems, urban redevelopment projects, and suburban shopping malls are pretty big too. Then there are skyscrapers, mega-projects, new-towns, edge-cities, again all large, obvious, written across the landscape and close to the heart of urban geography's sense of itself. And that is to say nothing of deindustrialization, suburban-sprawl, or inner-city dereliction and abandonment. Nor is it to mention the tens of thousands of miles of pipes, cables, and fibre optics that allow a contemporary city to function and which urban geography, led by innovate scholars like Stephen Graham and Simon Marvin (2001), has recently – if a little belatedly – discovered as being central to contemporary patterns of urbanization. Urban geography is also a discipline populated with Big Theories. Much of the most interesting

work that formed the so-called 'quantitative revolution' in human geography was urban. Melvin Webber (1964) was talking about the 'non-place urban realm' decades before the much trendier French anthropologist Marc Augé (1995) rediscovered the term in the 1990s, while Brian Berry with his central place hierarchies and rank size models was the world's most cited human geographer through much of the 1960s and 70s (see Berry, 1973). Indeed, Berry was toppled from this throne by another urban geographer with a Big Idea, David Harvey. A committed and intellectually charismatic Marxist who had begun his career as a quantitative geographer, Harvey (1973, 1982, 1989) argued that the 'urbanization of capital' was at the heart of the evolution of contemporary capitalism. According to him, if we want to understand how capitalism works we have to understand how cities work (see Castree, Chapter 8 this volume; Woodward & Jones, Chapter 15 this volume). More recently the UCLA-based urban geographer Ed Soja (1989, 1996, 2000) has been undertaking a hugely ambitious rethinking of the ontological foundations of human geography through the contemporary urban landscape (see Minca, Chapter 16 this volume). And, if that were not enough Big Ideas for one sub-discipline, Michael Dear (2000, 2002) has been trying to do something similar to Soja in a series of books that seek to trace the outline of a postmodern urbanism that does not merely mark out a new pattern of urbanization, but also the rudiments of a new postmodern social science.

Given its attraction to the Big, it is perhaps of little surprise that urban geography has not been particularly good at, or indeed often even interested in, making sense of many of the smaller elements that make up a city. The day-to-day routines of shopping and household provisioning, the life of public places like parks, sidewalks and shopping malls, the networks of friendship and enthusiasm, and so forth that give urban life so much of its texture have not often had a prominent place within the sub-discipline's theoretical or empirical imagination. It has by and large been urban sociologists, anthropologists and social historians, who have furnished us with these kinds of more intimate, more street level, narratives of the city. It is not just that urban geography has not been much good at making sense of these diverse everyday practices. There seems to be something in the quality of these practices that is resistant to the Big Theorizing dominant throughout urban geography. A key question for urban geography is whether this matters? Does urban geography need to be able to speak to these everyday practices? Is the texture of this intimate city, the 'sense of the marvellous suffusing everyday life' (Aragon, 1971: 23) that it speaks to, necessarily important to what urban geography is about? Could it simply be within the 'natural' division of the social sciences that urban geography should only try to speak of the big? That urban geography's natural terrain is the broad overview, the synthesizing, somewhat distanced, gaze? Or, does the inability to write about the everyday ecologies of cities represent something more significant? Does it symbolize a failure of the urban-geographical imagination that limits its usefulness, its applicability?

Ash Amin and Nigel Thrift's (2002) *Cities: Re-imagining the Urban*, is a bold and ambitious attempt to address – if not always directly answer – the above questions. In the space of just over 150 pages they seek to outline a set of novel ways of thinking about urban life, ways of thinking that transcend, or at the very least go beyond, the orthodoxies prevailing throughout both urban geography and urban studies. This might sound like another attempt at yet more Big Theory, the production of yet more Big Ideas. But Amin and Thrift's argument throughout *Cities* is more subtle than that. Urban geography and urban studies is, they argue, underpinned by a series of largely taken-for-granted, and apparently obvious, assumptions about just what elements make a city a city. Amin and Thrift want to place into question these underlying assumptions, and at the same time suggest an alternative framework through which to go about understanding the urban. They want to try to create a more open, more vital, and more populated style of urban analysis. A style of analysis that recognizes the ongoing incompleteness, the fuzzy-ness, the strange, often unpredictable, elements that make up a city. In Amin and Thrift's own words:

> The city has no completeness, no centre, no fixed parts. Instead, it is an amalgam of often disjointed processes and social heterogeneity, a place of near and far connections, a concatenation of rhythms; always edging in new directions. This is the aspect of cities which needs to be explained. (Amin and Thrift, 2002: 8)

What Amin and Thrift want to do is not simply demonstrate the importance of the so-called everyday. They want to try to steer us away from thinking in simplistic terms of big or small, near and far, global or local, and instead encourage us to explore how through all sorts of relationships cities emerge through complex patterns of connection. In short, they want to invite their readers to try and explore a different kind of urban *ontology*.

Key arguments: Toward a new ontology of the urban

How then should we go about developing this new kind of urban analysis? What tools do we need? What styles of thinking? For Amin and Thrift there are two key elements that need to be addressed in answer to these questions. The first question concerns the metaphorical repertoire through which urban geography and urban studies frames its understanding of cities. This needs to be enlarged and invigorated. Put simply, we need a new, and more inventive, set of metaphors in addition to those we are used to working with. Without new metaphors – and the shift in style of thinking that new metaphors bring with them – urban geography and urban studies will be unable to do proper justice to the heterogeneity and complexity that Amin and Thrift are pointing to. Secondly, they explore the contention that urban geography and urban studies needs to reach beyond the relatively narrow range of intellectual traditions upon which it had depended for much of its history. Drawing on a heterogeneous and often neglected collection of thinkers ranging from seventeenth and eighteenth-century European philosophers such as David Hume, John Lock and Baruch Spinoza, through to early twentieth-century writers such as the psychologist William James, the sociologist Gabriel Tarde, and the philosophers Henri Bergson and Alfred Whitehead, and later to the work of Michel Serres, Gilles Deleuze, and finally through to contemporary writers like the French philosopher Bruno Latour. Amin and Thrift (2002: 27) argue that we need to develop ways of considering cities that place stress on what they call 'an ontology of encounter or togetherness based on the principles of connection, extension and continuous novelty' .

This hardly sounds like we are clarifying matters any. But it is worth sticking a little longer with the intellectual inspirations animating Amin and Thrift's account because the whole sense of the argument that energizes *Cities* from the start of the book to the very finish depends on the distinctiveness of these inspirations. For Amin and Thrift, the analysis of cities that emerges through an engagement with the intellectual traditions mentioned above has a very distinctive style:

> All philosophies of becoming have a number of characteristics in common. One is an emphasis on instruments, on tools as a vital element of knowing, not simply as a passive means of representing the known. The second is their consideration of other modes of subjectivity than consciousness. The third is that 'feelings', howsoever defined, are regarded as crucial to apprehension. The fourth is that time is not a 'uniquely serial advance' … but rather exists as a series of different forms knotted together. Fifth, becoming is discontinuous, 'there is a becoming of continuity, but no continuity of becoming' … And finally, and most importantly, this means that 'prehensions' (ideas about the world) can constantly be built. More and more can be put into the world. (Amin and Thrift, 2002: 27–28; the quotes are from Whitehead, 1978)

So, Amin and Thrift want to think about ways of writing about cities that actively account for the role of all sorts of non-human actants in the life of cities. They want to explore ways of writing about the cities that actively engage with how they are inhabited by all kinds of bodies, non-human and human. They want to develop ways of writing about cities that recognize how cities are made (and *emerge*) through multiple, and often conflicting, orderings of time and space. And they want to try and find ways of writing about urban life that respect the complex, heterogeneous, becomings that define cities – ways of writing that pay attention to the many unlikely, surprising, and unexpected patterns of association that are constantly animating cities in all sorts of ways.

Key themes and content: networked, machinic, and powerful cities

To develop a sense – *feeling* might be a more appropriate word – of what an analysis of the urban organized through such an analytical style might look like, Amin and Thrift offer the reader five different accounts of city life in the substantive chapters that follow the book's first chapter. At the risk of some substantial simplification, the thematic concerns of these five chapters can be summarized under three main headings, developing ideas of cities as networked, as machinic and as 'powerful' (Amin and Thrift, 2002: 105).

In relation to the first of these, Amin and Thrift work through debates about the distanciation of social and economic relations, alighting on ideas that geographers need to think about cities as sites in distanciated networks. Of course, the idea of cities as networks is not new. The urban sociologist Manuel Castells (1989, 1996) has placed the idea of network at the centre of his analysis of informational cities for almost two decades. And well before that, researchers interested in urban community such as Barry Wellman (1979) and Claude Fischer (1982) had made extensive use of the term. Amin and Thrift, however, substantially extend, and reformulate, the network metaphor. For them the metaphor of the network is productive, not just because it speaks of patterns of connection, but because of how it suggests novel ways of considering how connections are built. In Amin and Thrift's thinking networks include not just humans and social institutions. They also include all sorts of non-human entities that allow a network to come into being. They include the computers, the communication systems, the software, the bureaucratic procedures, the meeting places; in short all the various assemblages of materials

that allow a network to achieve the stability it needs to be called a network. Thus, the networks they are talking about should more properly be called, following the work of writers like Bruno Latour (1993) and Michele Callon (1998), *actor*-networks.

Thought about in these terms, the metaphor of network is useful for Amin and Thrift for at least four reasons. Firstly, it is useful because it places into question both *who* and *what* is acting. It allows us to think about how social action is *distributed* across a range of entities rather than originating from an apparently autonomous individual. One of the most striking things about contemporary cities is the number of non-human actants that are monitoring, analyzing, and acting into the world. So, for example, Amin and Thrift (2002: 125) stress 'the modern city exists as a haze of software instructions'. These instructions form a kind of digital unconscious without which the modern city simply would not be able to function. A second strength of the network metaphor is that it helps to think about how boundary and scale are defined in a world characterized in all sorts of ways by distanciated relationships. Networks often generate topologies of connection that bear little relationship to Euclidian space. So, to highlight but one example, within a transnational corporation what counts as 'local' is that which is integrated into the organization – part of its communication structure and bureaucratic protocols – rather simply that which is simply nearby. Thirdly, the network metaphor is also useful because it highlights the knowledge, effort, skill and work that goes into making the world have the spatial-temporal feel – the rhythms – that it does. Amin and Thrift, thus, spend a great deal of time focusing on the amount of resources that go into the *maintenance* of the ordinary life of a city, and the precariousness of this achievement. They describe cities populated not just with high powered professionals, and global corporations, but ones that are also crucially dependent on all sorts of technicians, and service workers to keep them going. Fourthly, and finally, the metaphor of network is useful as it encourages us to think of cities as not primarily defined by proximity *per se* (as it is in so much of contemporary urban studies) but as 'sites' (Amin and Thrift, 2002: 63) for the staging of certain kinds of proximity.

Like the idea of the city as network, the metaphor of cities as machines also features large in *Cities*. The notion of cities as machines clearly has an established heritage within urban studies and urban geography (Mumford, 1934; Molotch, 1976). But again Amin and Thrift (2002: 78) have something distinctive in mind when they write of the need 'to understand the city as a *machine*':

> In using [machine] we do not mean to imply that the city can be understood through mechanical metaphors – with inputs, mechanisms and outputs – but rather as a 'mechanosphere', a set of constantly evolving systems or networks, machinic assemblages which intermix categories like the biological, technical, social, economic, and so on, with the boundaries of meaning and practice between the categories always shifting. (Amin and Thrift, 2002: 78)

As with the metaphor of the network, Amin and Thrift's use of the metaphor of the machine – and the associated pronoun, machinic – is productive because it points to how cities are built through the organizing of all sorts of materials and tools. The notion of cities as machines also stresses how all these materials can not only 'act', but also develop to generate all sorts of unlikely and unexpected new assemblages that exceed the origins plans, or patterns, of use intended for them.

So, the metaphor of the city as machine – at least in the sense Amin and Thrift use it – removes humans from being the *necessary* centre of any account of cities. Instead, they become but one element, or set of elements, albeit very important, along with all the other elements that make up the ecology of cities. This ethological approach introduces a whole new population of entities into urban geography. If we focus on cities as ecologies in the fullest sense then suddenly we are confronted by a world populated not just with purposeful actors, but also all kinds of different entities, ranging from viruses and bacteria, to buildings, rivers and deserts, to brown rats, cockroaches, and urban foxes – and this is to say nothing of the intricate technological assemblages already mentioned in the discussion of networks above. This ethological approach also prompts us to consider the need to understand ourselves as biological entities, as yet another species in the mechanosphere of cities, moved by all sorts of *passions*.

In adding passions to the metaphor of machine, Amin and Thrift are drawing direct inspiration from the French philosopher Gilles Deleuze's rereading of the seventeenth-century Dutch philosopher Baruch Spinoza. As Amin and Thrift (2002: 84) put it, Spinoza's 'thought can best be conceived as a kind of physics of bodies in which the human body is not a self-contained whole but built out of bodies with our own'. To think about passions, then, is also to think about the ways in which cities are made up of all sorts of communities organized through particular assemblages of passion. These can range from the intimate and emotionally intense such as with the connection of families, of lovers, of friends, to the communities of enthusiasm that include practices as diverse as 'Civil War re-enactment societies, the showing of cats and dogs, or trainspotting' (Amin and Thrift, 2002: 83) and fishing, to the assemblages of mutual purpose and indifference that are the crowded spaces of big city mass public transportation systems. And finally, to think about passions, and think about cities as machinic, is also to highlight how cities emerge out of the knotting together of a whole range of complex mechanisms for generating and regulating passions. Cities function through the constant and iterative 'engineering of certainty' (Amin and Thrift, 2002: 93) through all kinds of organizing mechanisms, from postal systems, and transportation networks, to the sports leagues and popular media that help lend the ebb and flow of a city's economy of passions a certain structure.

To think about the 'engineering of certainty' leads directly to the third and final overarching theme of *Cities*; cities as sites of power. In stressing the importance of network and flow, and parallel to that the place of 'passions', Amin and Thrift might seem to be outlining an overly cheery, voluntaristic, account of cities. To read their argument in these terms is to misinterpret what Amin and Thrift are trying to do. Throughout the pages of *Cities* they stress that cities are striated with power in all kinds of ways. However, for Amin and Thrift if the 'ontology of becoming' that they are outlining is to be treated seriously, then there is a need also to reconsider how the power that circulates in cities works. So, Amin and Thrift develop a distinctive approach to power:

> Rather than focusing on the conventional interest in urban studies on the domination and oppression by certain kinds of actors and institutions, our concern is with power as a mobile, circulating force which through the constant re-citation of practices, produces self similar outcomes moment by moment. (Amin and Thrift, 2002: 105)

This way of thinking through power is perhaps best understood as 'diagrammatic', in the sense that French historian Michel Foucault talked of forms of modern power and control being organized through certain 'diagrams' for considering how society should be assembled and governed. These 'diagrams of power' operate as 'abstract machines' (Amin and Thrift, 2002: 106) that work to process the world into a certain kind of consistency. Modern cities, then, might be seen as being organized through at least four diagrams: bureaucracy, production, sensuality, and imagination. Cities are permeated with a will to count, collect, categorize, and order all sorts of information and material. Equally, they are driven by a will to produce, whether that be economic growth, clean streets, or educated citizens. They are always organized around tasks of produce, either more of something, or in some cases less. Rarely are they organized around keeping things exactly as they are. Similarly, cities generate and organize all sorts of sensualities, and just as they generate them, so do they almost simultaneously throw up all sorts of techniques and technologies to manage, regulate, and control them. And, lastly, cities are remarkable fields for imaginative endeavour. The shaping and regulating of this imaginative space is central to the exercise of any modern organization.

Reception

Cities is not an easy book to digest. It is designed to challenge and question the reader's view of what cities are and how they should be understood. This wilful awkwardness has shaped much of *Cities*' critical reception to date. On the one hand, in the short time since its publication the book has been widely read and highly cited both within urban geography, urban studies, and indeed elsewhere in the social sciences. This is hardly surprising, given the prominent place both Ash Amin and Nigel Thrift had already established for themselves within urban geography and urban studies by the time of *Cities*' publication: between them Amin and Thrift had published dozens of articles and book chapters on a wide range of urban issues, and both held professorships at prominent British geography departments (Amin at Durham University, Thrift at the University of Bristol). As such, *Cities* has been read by many as simultaneously a kind of summary of Amin and Thrift's key arguments, and an overview (following Thrift, 1996) of how one might go about doing a so-called 'non-representational' style of urban geography.

On the other hand, it is not at all clear how sustained urban geography and urban studies' engagement has been with the arguments made in *Cities*. While most of the reviews of the book in academic journals were broadly positive, almost all raised doubts about how a researcher would actually go about *doing* Amin-and-Thrift-ian inspired urban research. In part these doubts were because, as one sympathetic reviewer wrote (Savage, 2003: 807), its 'perspective is so distinctive that it does not engage with much "actually existing" urban scholarship'. It follows that to get more than just a passing sense of the kind of accounts that Amin and Thrift are asking urbanists to try to write requires the reader to invest an enormous amount of intellectual energy going back to writers like Bruno Latour, Gilles Deleuze, and Alfred Whitehead through which the argument of *Cities* is developed. And while, as Amin (2007) shows in a recent article surveying urban geography, there *is* a growing body of literature that follows the lines of thinking outlined in *Cities*, this writing has hardly come to define the mainstream of the sub-discipline.

More tellingly, many reviewers questioned the applicability of Amin and Thrift's new style of urban theory. These questions were both conceptual and methodological. Some wondered what kinds of methods and techniques researchers would need to use to operationalize Amin and Thrift's theoretical vision. Would they simply use those they were already using? Or were a whole bunch of new methods required? Or some mix of the old and the new? These are questions that, as Ian Gordon (2003: 519) wrote in a review in *Progress in Human Geography*, 'are never really confronted'. Gordon also wondered why it was that – in his reading – 'all

notion of inequalities of interest and power disappear' (Gordon, 2003: 520) in Amin and Thrift's account. For all the interesting and novel elements of urban life that *Cities* throw up, it seems to miss what many urban researchers would consider the 'brute facts about urban life' (Gordon, 2003: 520). That is to say, it somehow loses a sense of all the ways in which cities are organized around questions of life and death; how they are defined by profound and often deep-seated poverty and exclusion, how they are often sickeningly unequal and unfair places. As Lynn Staeheli (2004: 86), put it in her review in *Urban Geography* '[i]n this time of the revanchivist city, the "soft" politics implied in [*Cities*] seem somehow not enough'.

Working with Amin and Thrift

Populated with urban anglers, foxes, streetlights, airports, computer software, networks of friendship and enthusiasm, computer screens and keyboards, hands and fingers, roller-coasters, bacteria and viruses (microbiological and electronic), lawns, automobiles, databases and accounting systems, films and novels and guidebooks, regulations of all different kinds, telephone networks, styles of talk, league tables, all sorts of non-human and human passions, *Cities: Re-imagining the Urban* presents a striking, distinctive, and intellectually challenging attempt to re-think urban geography and urban studies. It also offers a succinct introduction to the intellectual projects of two of the most important urban geographers writing today.

Nonetheless, *Cities* is more than just a summary of arguments that Amin and Thrift had outlined in previous publications. As Amin and Thrift write at the start of the book:

> We see [*Cities*] as a kind of staging post towards a different practice of

urban theory based on the transhuman rather than the human, the distanciated rather than the proximate, the displaced rather than the placed, and the intransitive rather than the reflexive. (Amin and Thrift, 2002: 5)

Cities, then, is much more about outlining a project in the making than a summary of a well-established intellectual position. This makes for exciting reading. Throughout *Cities* one often has little sense of where Amin and Thrift will go next in their argument. And equally, one is also constantly surprised by the material that they draw on to elaborate their argument. Do they really think that a history of lawns would tell us a great deal about contemporary patterns of urbanization? Do they really think that we should carefully consider the sociality of urban anglers? That we need to be aware of the conversational rhythms of telephone conversations?

Well, the obvious answer is, 'Of course they do!' The real question the reader needs to ask after reading through *Cities* is not, 'Do Amin and Thrift's examples hold up?' It is, 'Does *Cities* outline an approach to thinking about cities that actually works?' And, along with that question, the reader must also ask, 'Do Amin and Thrift offer us *enough* of a sense of what this new style of urban analysis, this study of cities as sites of constant *becoming*, involves?' Which leads to a third, and equally important, question, 'Do Amin and Thrift manage to map out a way of thinking about cities that actually offers urban geography and urban studies more analytical leverage than the theoretical approaches that it already has?' Well, the answer to each of those questions, unfortunately, is that it is too early to tell. Amin and Thrift's book is too short, and, the enormous productivity of each of the co-authors aside, not nearly enough material has been published elsewhere on cities in the style that *Cities* is arguing for to really make a careful judgement on any of these three questions.

Conclusion

Ultimately, *Cities* – as all new ways of thinking are – is dependent on a kind of wager. A wager that in working to show up all sorts of novel elements of urban life, that in trying to think about cities in a style outside of the established grooves of conventional urban geography and urban studies, it will invigorate and expand our understanding of how cities can be thought about in ways that existing approaches simply cannot. In a sense, then, one can only discover if the ideas animating *Cities* work by trying to work with them and see what one comes up with. So, whether *Cities* will cure urban geography and urban studies of its love of Big Things and Big Ideas is perhaps doubtful. But, read with the sense of intellectual adventure with which Amin and Thrift approached its writing, *Cities* is one of those surprisingly rare books within human geography, a book that bears repeated study.

Secondary sources and references

Amin, A. (2007) 'Re-thinking the urban social', *City* 11 (1): 100–114.

Amin, A. and Thrift, N. (2002) *Cities: Re-imagining the Urban*. Oxford: Polity.

Aragon, L. (1971) *Paris Peasant*. London: Picador.

Augé, M. (1995) *Non-Places: Introduction to the Anthropology of Supermodernity*. London: Verso.

Benjamin, W. (1973) *Charles Baudelaire*. London: Verso.

Benjamin, W. (1978) *One Way Street and Other Writings*. London: Verso.

Berry, B. (1973) *The Human Consequences of Urbanisation: Divergent Paths in the Urban Experience of the Twentieth Century*. London: Macmillan.

Callon, M. (1998) *Laws of Markets*. Oxford: Blackwell.

Castells, M. (1989) *The Informational City*. Oxford: Blackwell.

Castells, M. (1996) *The Network Society*. Oxford: Blackwell.

Caygill, H. (1998) *Walter Benjamin: The Colour of Experience*. London: Routledge.

Dear, M. (2000) *The Postmodern Urban Condition*. Oxford: Blackwell.

Dear, M. (ed.) (2002) *From Chicago to LA: Making Sense of Urban Theory*. Beverley Hills: Sage.

Fischer, C. (1982) *To Dwell Among Friends*. Berkeley: University of California Press.

Gordon, I. (2003) 'Review of Amin and Thrift *Cities: Re-imagining the Urban*', *Progress in Human Geography* 27 (4): 519–520.

Graham, S. and Marvin, S. (2001) *Splintering Urbanism*. London: Routledge.

Harvey, D. (1973) *Social Justice and the City*. Oxford: Blackwell.

Harvey, D. (1982) *The Limits to Capital*. Oxford: Blackwell.

Harvey, D. (1989) *The Urban Experience*. Oxford: Blackwell.

Latour, B. (1993) *We Have Never Been Modern*. Hassocks: Harvester Wheatsheaf.

Lefebvre, H. (1991) *The Production of Space*. Oxford: Blackwell.

Lefebvre, H. (2004) *Rhythmanalysis: Space, Time and Everyday Life*. London: Continuum.

Molotch, H. (1976) 'The city as a growth machine', *The American Journal of Sociology* 82: 309–318.

Mumford, L. (1934) *Technics and Civilisation*. New York: Harcourt and Brace.

Savage, M. (2003) 'Review of Amin and Thrift *Cities: Re-imagining the Urban*', *Sociology* 37 (4): 806–808.

Soja, E. (1989) *Postmodern Geographies*. London: Verso.

Soja, E. (1996) *Third Space*. Oxford: Blackwell.

Soja, E. (2000) *Postmetropolis*. Oxford: Blackwell.

Staeheli, L. (2004) 'Review of Amin and Thrift *Cities: Re-imagining the Urban*', *Urban Geography* 25 (1): 84–86.

Thrift, N. (1996) *Spatial Formations*. London: Sage.

Thrift, N. (2005) *Knowing Capitalism*. London: Sage.

Webber, M. (1964) *Explorations into Urban Structure*. Philadelphia: University of Pennsylvannia.

Wellman, B. (1979) 'The community question: the intimate networks of East Yorkers', *American Journal of Sociology* 84, March: 1201–1231.

Whitehead, A.N. (1978) *Process and Reality*. New York: Free Press.

26 FOR SPACE (2005): DOREEN MASSEY

Ben Anderson

For the future to be open, space must be open too. (Massey, 2005: 12)

Introduction

On page 108 of *For Space,* an impassioned book that discloses the theoretical and political challenges of thinking space, a map of part of the South-East of England is inscribed with a very simple if perhaps initially puzzling phrase: 'ceci n'est pas l'espace'

The phrase recalls Rene Magritte's famous inscription below a painting of a pipe: 'ceci n'est pas une pipe' (*this is not a pipe*). Initially, like Margritte's phrase, it may seem odd – counterintuitive perhaps – since we are being rather bluntly informed that a map of roads

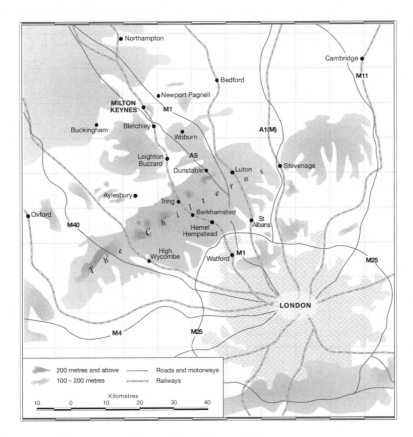

Figure 26.1 'Ceci n'est pas l'espace' (Figure 11.1 in Massey, 2005)

and motorways, railway lines, topography, fields and villages is not space. Odd because maps have become central to how we think about and imagine space. Yet maps, perhaps those we are most familiar with, function by representing space as an ordered surface in relation to which the observer is positioned outside and above. Massey's point is a simple one that is now echoed in a critical literature on cartography – that hegemonic types of mapping represent space as a 'completed horizontality' – in which the dynamism of change is exorcised in favour of a totality of connections. Mapping is one of a number of ways in which the disruptiveness of space is tamed. Offering an alternative non-euclidean imagination of space, that disrupts this and other problematic accounts of space, is therefore the pressing task that animates *For Space*: a book that Massey (2005: 13) summarizes as comprising 'an essay on the challenge of space, the multiple ruses through which that challenge has been so persistently evaded, and the political implications of practising it differently'.

The basis to an alternative approach to space can be articulated in a set of three intertwined propositions:

- Space is the product of *interrelations*; thus we must recognize space 'as constituted through interactions, from the immensity of the global to the intimately tiny' (Massey, 2005: 9).
- Space is the sphere of the possibility of the existence of *multiplicity*; that is space 'as the sphere in which distinct trajectories coexist; as the sphere therefore of coexisting *heterogeneity*' (Massey, 2005: 9).
- Space is always *under construction*; 'it is always in the process of being made. It is never finished; never closed' (Massey, 2005: 9).

For Space is an argument for the recognition of these three characteristics of space

and for a lively, heterogeneous, progressive politics that thereafter responds to them. The three propositions therefore aim to enable us to ponder the challenges and delights of spatiality and subsequently open up the political to the challenge of space – perhaps disrupting how political questions are formulated, perhaps intervening in current arguments and perhaps contributing to alternative imaginations that enable different spaces to be.

The double aim of *For Space* – to simultaneously open up our thinking of the spatial and the political – resonates with Massey's work over the past two decades. From research on industrial restructuring and the social division of labour (see Massey, 1984; Phelps, Chapter 10 this volume), through to theoretical work on the emergence and disruption of power-geometries (see Massey, 1994), Massey has been a consistent advocate of the political necessity of teasing out the mutual imbrications of the spatial and the political. If *For Space* therefore chimes with several of Massey's abiding concerns then it also resonates with the emergence of a range of poststructuralist geographies that associate space with dynamism and thus qualities of openness, heterogeneity and liveliness (see, for example, Amin and Thrift, 2002; Doel, 1999; Murdoch, 2006; Whatmore, 2002). The other context she writes in is, however, the persistence of a set of problematic associations around space that we have inherited from a set of philosophical lineages and that are constantly articulated in contemporary politics. The first section of this essay reviews, therefore, Massey's critical engagement with other imaginations of space. Section two moves on to draw out the alternative conception of space that *For Space* outlines by returning to explicate the three propositions introduced briefly above. Section three then thinks through more precisely how Massey's alternative conceptualization of space offers and promises Human Geography a type of 'relational politics'. In the conclusion I raise a series of questions about the relational approach to space that *For Space* exemplifies and argue that what is distinctive about the book is that it offers a specific ethos of engagement which trusts that 'there are always connections yet to be made, juxtapositions yet to flower into interaction (or not, for not all potential connections have to be established), relations which may or may not be accomplished' (Massey, 2005: 11).

'Unpromising associations'

The title of Doreen Massey's book, *For Space*, provokes a simple question. Why *For Space*? The title declares that space *matters*. That it inflects how we engage, understand and approach the world. So conceptualizing space should, therefore, be a pressing concern for us – it should cause us problems, make us think, and interest us. Yet the title is not *about* space or *thinking* space or *questioning* space. By declaring she is *for* space Massey affirms the possibilities and potentialities enabled by space(s). I will interrogate these possibilities in sections three and four but before we can disclose them we need to interrogate the 'unpromising associations' that, for Massey, serve to conceptualize or assume space to be simply the negative opposite of time. Despite the reassertion of space in social theory which has made space part of the lexicon of the social sciences and humanities over the past two decades or so, deeply ingrained habits of thought continue to tie space to a set of dehabilitating assumptions. These are assumptions that are fundamentally embedded in the framing of a range of contemporary problems. Central to the history of modernity, for example, has been a translation of spatial heterogeneity into temporal sequence. Different places are interpreted as occupying different stages in a single temporal sequence in the

various stories of unilinear progress that define the West against the rest (such as modernization or development). Talk of the 'inevitability' of neo-liberal 'globalization', to give another example, assumes both a free unbounded space and that globalization takes only one form. In both cases – and we can think of others such as the idea that space can be annihilated by time – the contemporaneous heterogeneity of the world is all too easily forgotten and thus difference erased.

In aiming to discern how such a taming of the spatial is also present in a range of philosophers and political theorists, Massey's concern is not, it should be noted, simply with how time has been prioritized over space – a claim that has been central to the reassertion of space but is itself tied to problematic assertions that we live in uniquely 'spatial times' (e.g. Soja, 1989). Instead she interrogates how space has been attached to a set of 'unpromising associations' in the work of a set of theorists and theories broadly understood as either structuralist or post-structuralist (including Althusser, Bergson, Laclau and Derrida). She describes her relation with these theorists and schools of thought in strikingly affective terms. In relation to their treatment of space she is:

> Puzzled by a lack of explicit attention they give, irritated by their assumptions, confused by a kind of double usage (where space is the great 'out there' and the term of choice for characteristics of representation, or of ideological closure), and, finally, pleased sometime to find the loose ends (their own internal dislocations) which make possible the unravelling of those assumptions and double usages and which, in turn, provokes a reimagination of space which might be not just more to my liking, but also more in tune with the spirit of their own enquiries. (Massey, 2005: 18)

Despite her puzzlement and irritation, the last line in this quote stresses that Massey's engagement with this range of thinkers is a reparative rather than dismissive one. Rather than condemn them, and in that act of dismissal separate her own approach from theirs, Massey's critique aims to disclose a range of new potential openings. Each of the theorists, and schools of thought, offer something to Massey's project. From Bergson she understands questions of the dynamism of life – of liveliness. Structuralism offers an understanding of how the identity of entities is made out of relations; whilst deconstruction heralds a constant enlivening interruption to space. Yet in her engagement with each she argues that space takes on a set of two 'unpromising associations' that either implicitly or explicitly tame space and refuse the challenge of understanding its singularity as the realm of 'radical contemporaneity'. First, a conceptualization of space as *static* that equates space with a stabilization of life. Space is assumed to conquer the inherent dynamism of time by imposing an order upon the life of the real – 'spatial immobility triumphs temporal becoming' (Massey, 2005: 30). Second, a conceptualization of space as *closed* and thus awaiting the enlivening effects of temporality for change or anything new to take place. Instead then of thinking space as the very condition of and for radical contemporaneity, that is the sphere of co-existing multiplicity, space is tied to the chain stasis/closure.

Alternatives

It is because of the promise of space, that is what it could offer us or may give us, that Massey critiques the unpromising associations that, firstly, casts space as separate from time and then, secondly, devalues space by making it the negative opposite of time. In other words, her engagement with theorists, and schools of thought, is animated by a belief that imagining an alternative understanding of space is a pressing intellectual task because it is simultaneously a means of responding to spatial politics. This task is therefore not only to critique taken-for-granted uses of space but to offer alternative conceptualizations that could help the difficult work of building alternatives to various 'power-geometries' – including neoliberal globalizations. Massey's positive alternative conceptualization of space can be placed in the context of a range of diverse engagements that think space and place in terms of relationality (i.e. where relations, types of connection or association between entities, precede identity). Such a move resonates with a set of trajectories in human geography that no longer conceptualize space as a 'container' in which other entities or processes happen. Instead, any space or place, from the intimate space of a body to the space of the globe, are precarious achievements made up of relations between multiple entities. Spaces have to, in other words, be made and remade because relations are processual. A named space, such as London or Newcastle, does not have a permanent essence.

Relational thought takes a number of quite different forms in Human Geography. Harvey (1996), in advocating a type of dialectical materialism influenced by a long lineage of process thinking, argues that space is made by (biological, physical, social, cultural) processes and that these processes are themselves constituted by relations between very different kinds of entities. Thrift (1996), advocating a 'modest' style of theory that he terms non-representational, conceptualizes space as a site of becoming that has to be constantly performed in and through numerous everyday practices. There is much that Harvey and Thrift disagree on, but what enables them both to be cast, like Massey, as relational thinkers is that discrete spaces and places are permanencies that are only ever provisionally stabilized because of the multitude of entities in relation that they are constituted from.

For Space is perhaps the most detailed statement of an approach cast in terms of relations, and relationality, so it is important to pause and unpack in more detail the three propositions that make up the core features of Massey's alternative approach. First, in concert with the claims of relational thought, Massey (2005: 107) argues that space is constituted *through* its relations. Outside of these relations a space has no existence. There is no difference here between spaces we would, ordinarily, consider to be 'big' or 'small'. All are products of relations between all manner of heterogeneous bits and pieces (that are simultaneously natural, social, political, economic and cultural). Space is thus a sphere 'of dynamic simultaneity, constantly disconnected by new arrivals, constantly waiting to be determined (and therefore always undetermined) by the construction of new relations. It is always being made and is always therefore, in a sense, unfinished (except that "finishing" is not on the agenda)'. This means that, secondarily, space is the sphere of multiplicity because it is made out of numerous heterogeneous entities. Space is the gathering together of multiple openended, interconnected, trajectories to produce what Massey (2005: 111) terms that 'sometimes happenstance, sometimes not – arrangement-in-relation-to each-other'. This multiplicity means that space is the condition for the unexpected. Third, and consequently, space is an ongoing achievement that is never finished or closed. Stabilities and permanencies, a place that appears unchanging, for example, are provisional achievements that have to be constantly made and remade (even if this process of making and remaking is hidden or taken-for-granted).

An example that Massey uses that exemplifies how these three propositions function together to disclose space differently is an example of a train journey from London to Milton Keynes. In a journey you are not simply travelling *through* space or *in* space (that is from one named place – London – to another – Milton Keynes). This would make space into a simple container within which other things only happen. Instead you minutely alter it – if only a little bit by virtue of your presence in one place and your absence from the other place – and thus contribute to its being made. Yet as space is altered – by your active material practices – the places are themselves constantly moving on and changing as they are constituted out of processes that exceed you:

> At either end of your journey, then, a town or city (a place) which itself consists of a bundle of trajectories. And likewise with the places in between. You are, on that train, travelling not across space-as-a-surface (this would be the landscape – and anyway what to humans may be a surface is not so to the rain and may not be so either to a million micro-bugs which eave their way through it – this 'surface' is a specific relational production), you are travelling across trajectories. That tree which blows now in the wind out there beyond the train window was once an acorn on another tree, will one day hence be gone. That field of yellow oil-seed flower, product of fertiliser and European subsidy, is a moment – significant but passing – in a chain of industrialised agricultural production. (Massey, 2005: 119)

Human geography and a relational politics

From this evocative image of spaces emerging, and passing away, during a train journey we get a sense of the delight, or perhaps even wonder or joy (see Bennett, 2001), that Massey fosters as she carefully composes her alternative conceptualization of space and place as relational and thus fundamentally open. Another example she returns to is the place of Keswick – a town in the Lake District, UK – a town that is bound to the romance of the timelessness of the hills, a pre-given collective identity (based on a type of farming) and now modern practices of tourism. Using the case of a visit to Keswick by her and her sister, Massey argues that what is special about this place, and all others, is its 'throwntogetherness' – the way that very diverse elements that cross categories such as the natural or social come together to foster a particular 'here and now'. This is what makes places specific – this gathering of diverse entities into relation:

> This is the event of place. It is not just that old industries will die, that new ones may take their place. Not just that Hill farmers round here may one day abandon their long struggle, nor that that lovely old greengrocers is now all turned into a boutique selling tourist bric-a-brac. Nor, evidently, that my sister and I and a hundred other tourists soon must leave. It is also that the hills are rising, the landscape is being eroded and deposited; the climate is shifting; the very rocks themselves continue to move on. The elements of this 'place' will be, at different times and speeds, again dispersed. (Massey, 2005: 140/141)

In the example of Keswick as a particular place, and of the train journey as a type of movement, we see how the three propositions foster a shift in how we think about and encounter space – a shift announced in a proposition that Marcus Doel (2000) makes: echoing Massey and drawing on a range of poststructuralist thought he argues that 'it would be better to approach space as a verb rather than a noun. *To space* – that's all. Spacing is an action, an event, a way of being'.

For Space can, therefore, be read as attempt to think space as a verb – a move that ties

space to a set of problematics that have been seen as the provenance of time. How to think through the emergence of new spaces and places? How to live with difference within spaces and places? How to engage with the interconnections that tie together what we may consider to be 'separate' spaces and places? Space becomes, therefore, the very ground of the political because to think spatially is to engage with the existence of multiple processes of coexistence. That is, it opens up a type of relational politics based on the 'the negotiation of relations, configurations' (Massey, 2005: 147).

What is at stake is how politics makes a difference from within 'the constant and conflictual process of the constitution of the social, both human and nonhuman' (Massey, 2005: 147). How would a relational politics disclose and intervene in the constellation of trajectories that produce particular places or spaces? Massey offers three practices that follow from opening up the political to the spatial – that is to 'the challenge of our constitutive interrelatedness' (Massey, 2005: 195). First, a politics of receptivity that is open to the 'throwntogetherness' of place – the way that a place is 'elusive' because it is made out of multiple trajectories. Thus a politics of place would not be simply a politics of 'community' but would involve processes of 'negotiation' that would confront the fact of difference via 'the range of means through which accommodation, anyway always provisional, may be reached or not' (Massey, 2005: 154). The key, though, is that there are no portable rules because of the uniqueness of place: 'the negotiation will always be an invention; there will be need for judgement, learning, improvisation' (Massey 2005: 162). Second, and following on, there can be rules of space and place that cosily determine a political position, i.e. no spatial principles from which a position is simply deduced. Take, for example, arguments about the 'openness' of particular spaces. These are frequently fraught with contradiction. So those on the right of the political spectrum may argue for the free movement of capital but against the free movement of labour, whilst those on the left may argue for the free movement of people but against unbridled free trade. As Massey (2005: 166) stresses 'abstract spatial form, as simply a topographic spatial category, in this instance openness/ closure, cannot be mobilised as a universal topography distinguishing left and right'. The key instead is to think through the relations through which the spaces, and thus different types of openness and closure, are constructed without privileging a-priori either openness, movement and flight, or closure, stasis and immobility. Openness is not the same in the case of the free movement of capital as it is in the free movement of people. Third, if a relational politics requires both negotiations due to 'throwntogetherness' and a politics of the terms of openness and closure, it also requires a politics of connectivity that takes account of wider spatialities of relations. The fact of connectivity raises a host of difficult questions about responsibilities that it is the task of a spatial politics to open up:

> It questions any politics which assumes that 'locals' take all decisions pertaining to a particular area, since the effects of decisions would likewise exceed the geography of that area; it questions the predominance of territorially based democracy in a relational world; it challenges an all-too-easy politics which sets 'good' local ownership automatically against 'bad' external control (Amin, 2004). It raises the issues of what might be called the responsibilities of the local: what, for instance, might be the politics and responsibilities towards the wider planet of a world city such as London? (Massey, 2005: 181)

To finish with a set of open questions is therefore appropriate because what is promised by a relational politics is an expansion of the problems that animate 'the political'. This is an expansion that is energized not by the laying out of a set of invariant principles but by the gradual emergence of a distinctive style or ethos of engagement with the world: an ethos that strives to be attentive to the consequences of our varied interrelatedness. It therefore resonates with other current attempts to foster geographical imaginations that engage the world differently in and through relational imaginations of space. Whatmore (2002), animated by a range of non-representational theories, argues for an ethos of generosity that would enable us to understand the complex entanglements that fold humans and non-humans into specific 'hybrid' geographies. Gibson-Graham (2006), carefully sketching a post-capitalist politics, offer a hopeful stance that would disclose the relations that foster spaces of hope in order to disrupt the mastery of neoliberal capitalism. By resonating with these and other shifts in geographic thought and practice, Massey (2005) offers a means of thinking through politics of interrelations that is sensitive to heterogeneity of space and thus the genuine openness of the future, i.e. the very condition of the political.

Such an ethos of engagement with the world emerges from a positive understanding of space based simply on 'a commitment to that radical contemporaneity which is the condition of, and the condition for, spatiality' (Massey, 2005: 15). It therefore achieves two effects. On the one hand the relational alternative disrupts many of the taken-for-granted understandings about the relation between space and time that have a hold over the popular and political imagination and are also still played out by theorists that geographers are otherwise happy to encounter. Massey discloses an evasion of space and is sensitive to the ideological and hegemonic work that an association between space and the closed, immobile and fixed does. On the other hand, a relational approach to space fosters the emergence of a new set of questions that force us to wonder again about the task of spatial thought. Massey constantly discloses how thinking space fosters a commitment to radical contemporaneity. These two effects combine to open up the political to the challenge of space and thus disclose a host of new political questions and problems and therefore, perhaps, the faint outline of a geography based on practices of relationality, a recognition of implication and a modesty of judgement.

Conclusion

For Space exemplifies what a relational approach to theorizing space and place both offers and promises the ethos and politics of contemporary human geography. There are, therefore, a set of questions about relations and relationality that are emerging in human geography that may become central to how *For Space* is critiqued, evaluated and incorporated into the geographical imagination.

- On the one hand how do we understand the term 'relation' given that there are many forms of 'elation' (such as encounter, belonging, etc.). On the other, how do we understand relations of non-connection – what we could term 'non-relations'?
- How to understand the durability of particular places or spaces? How do certain constellations of relations repeat and endure? Alternatively, how to disclose those space times that flicker out of existence or those space times that never came to be?
- How to understand differences in spaces based on size, i.e. how to theorize scale from a relational and thus non-Euclidean perspective?
- How to engage in differences in degree and in kind within and between the entities that make and are made by relational spaces, i.e. how are the capacities to act of a human different to the capacities to act of a non-human?
- How to engage with radical alterity from within a system of relational thought. That is how to engage with relations that remain unknowable, undecided or indeterminate?
- How to engage with other types of spaces that Human Geography is only beginning to encounter – such as spaces constituted through the circulation of images or spaces animated by the distribution of affect – or the multiple topological forms that relational space can take (network spaces, Euclidean spaces, fluid spaces, etc.)?

Secondary sources and references

Amin, A. and Thrift, N. (2002) *Cities: Re-imagining the Urban.* London: Polity Press.

Bennett, J. (2001) *The Enchantment of Modern Life: Attachments, Crossings, and Ethics.* Princeton and Oxford: Princeton University Press.

Crang, M. and Thrift, N. (2000) *Thinking Space.* London: Routledge.

Doel, M. (1999) *Poststructuralist Geographies: The Diabolical Art of Spatial Science.* Edinburgh: Edinburgh University Press.

Doel, M. (2000) 'Un-Glunking geography. Spatial science after Dr Seuss and Gilles Deieuze', in M. Crang and N. Thrift (eds) *Thinking Space.* London and New York: Routledge, pp. 117–135.

Gibson-Graham, J.-K. (2006) *A Postcapitalist Politics.* Minnesota: University of Minnesota Press.

Harvey, D. (1996) *Justice, Nature and the Geography of Difference.* Oxford: Blackwell.

Massey, D. (1984) *Spatial Divisions of Labour.* London: Macmillan.

Massey, D. (1994) *Space, Place and Gender.* Cambridge: Polity Press.

Massey, D. (2005) *For Space.* London: Sage.

Murdoch, J. (2006) *Post-Structuralist Geography.* London: Sage.

Soja, E. (1989) *Postmodern Geographics: The Reassertion of Space in Critical Social Theory.* London: Verso.

Thrift, N. (1996) *Spatial Formations.* London: Sage.

Whatmore, S. (2002) *Hybrid Geographies.* London: Sage.

Index